Success in Math
Basic Algebra

Teacher's Resource Manual

Executive Editor: Barbara Levadi
Market Manager: Sandra Hutchison
Senior Editor: Francie Holder
Editors: Karen Bernhaut, Douglas Falk, Amy Jolin
Editorial Assistant: Kris Shepos-Salvatore
Educational Consultant: Kathleen Coleman
Production Manager: Penny Gibson
Production Editor: Walt Niedner
Interior Design: The Wheetley Company
Electronic Page Production: The Wheetley Company
Cover Design: Pat Smythe

Copyright ©1996 by Globe Fearon Education Publisher, a Division of Simon and Schuster, 1 Lake Street, Upper Saddle River, New Jersey 07458. All rights reserved. No part of this book may be reproduced or transmitted in any form or by any means electrical or mechanical, including photocopying, recording, or by any information storage and retrieval system, without permission in writing from the publisher.

Permission is given for individual classroom teachers to reproduce the blackline masters for classroom use. Reproduction of these materials for an entire school system is strictly forbidden.

Printed in the United States of America   2 3 4 5 6 7 8 9 10 99 98 97 96

ISBN 0-8359-1189-6

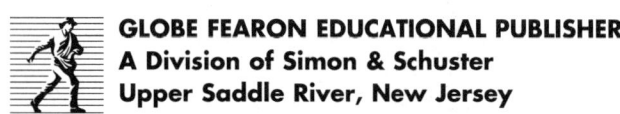

Contents

English Proficiency Strategies	v
Cooperative Learning	vi
Manipulatives	vii
Problem Solving	viii

Chapter 1 The Language of Algebra — 1
- 1•1 Understanding Variables and Expressions — 1
- 1•2 Understanding Exponents — 1
- 1•3 Using Grouping Symbols — 1
- 1•4 Understanding Order of Operations — 2
- 1•5 Using Inverse Operations — 2
- 1•6 Verbal Expressions and Algebraic Expressions — 2
- Chapter 1 Test — 3

Chapter 2 The Rules of Algebra — 4
- 2•1 Recognizing Like Terms and Unlike Terms — 4
- 2•2 Understanding the Properties of Numbers — 4
- 2•3 Understanding the Distributive Property of Multiplication Over Addition — 5
- 2•4 Adding Like Terms — 5
- 2•5 Subtracting Like Terms — 6
- 2•6 Multiplying Terms — 6
- 2•7 Dividing Terms — 6
- Chapter 2 Test — 7

Chapter 3 Equations and Formulas — 8
- 3•1 What is an Equation? — 8
- 3•2 Subtracting to Solve Equations — 8
- 3•3 Adding to Solve Equations — 9
- 3•4 Dividing to Solve Equations — 9
- 3•5 Multiplying to Solve Equations — 9
- 3•6 What Is a Formula? — 10
- 3•7 Focusing on Equations with Fractions and Mixed Numbers — 10
- Chapter 3 Test — 11

Chapter 4 Integers — 12
- 4•1 What Is an Integer? — 12
- 4•2 Integers and the Number Line — 12
- 4•3 Absolute Value — 12
- 4•4 Which Integer is Greater? — 13
- 4•5 Adding Integers — 13
- 4•6 Subtracting Integers — 13
- 4•7 Multiplying Integers — 14
- 4•8 Dividing Integers — 14
- Chapter 4 Test — 15

Chapter 5 Introduction to Graphing — 16
- 5•1 The Coordinate Plane — 16
- 5•2 Ordered Pairs — 16
- 5•3 Tables and Graphs — 16
- 5•4 Special Graphs — 16
- 5•5 Slope — 17
- 5•6 Finding the Intercepts of an Equation — 17
- Chapter 5 Test — 18

Chapter 6 More About Graphing — 19
- 6•1 Slope as Rate of Change — 19
- 6•2 Using the Slope-Intercept Form of an Equation — 19
- 6•3 Other Forms of Equations — 20
- 6•4 Graphing Absolute-Value Equations — 20
- 6•5 Other Graphs and Their Equations — 20
- Chapter 6 Test — 21

Chapter 7 Exponents — 22
- 7•1 More About Exponents — 22
- 7•2 Writing Very Large and Very Small Numbers — 22
- 7•3 Multiplication Properties of Exponents — 22
- 7•4 Division Properties of Exponents — 23
- 7•5 Negative Exponents — 23
- 7•6 The Exponent of Zero — 23
- 7•7 Radicals — 24
- Chapter 7 Test — 25

Chapter 8 Inequalities — 26
- 8•1 Equations and Inequalities — 26
- 8•2 Solving One-Step Inequalities — 26
- 8•3 Solving Two-Step Inequalities — 27
- Chapter 8 Test — 28

Chapter 9 Monomials and Polynomials — 29
- 9•1 Recognizing Monomials and Polynomials — 29
- 9•2 Adding and Subtracting Polynomials — 29
- 9•3 Multiplying a Polynomial by a Monomial — 30
- 9•4 Multiplying Polynomials — 30
- Chapter 9 Test — 31

Chapter 10 Factoring — 32
- 10•1 Finding the Greatest Common Factor — 32
- 10•2 Factoring Expressions — 32
- 10•3 Factoring Quadratic Expressions — 33
- 10•4 Factoring Differences of Two Squares and Perfect Square Trinomials — 33
- 10•5 Zero Products — 34
- 10•6 Solving Equations by Factoring — 34
- 10•7 The Quadratic Formula — 34
- Chapter 10 Test — 35

Chapter 11 Systems of Equations — 36
- 11•1 What Is a System of Equations? — 36
- 11•2 Solving a System of Equations by Graphing — 36
- 11•3 Solving a System of Equations Using Addition or Subtraction — 37
- 11•4 Solving a System of Equations Using Multiplication — 37
- 11•5 Solving a System of Equations Using Substitution — 38
- Chapter 11 Test — 39

Answers — 40

English Proficiency Strategies

Mathematics is a language consisting of carefully defined symbols that represent fundamental concepts. Students learning English as a second language (ESL) or limited in English proficiency (LEP) may need more time to make connections between concepts and mathematical language. All students in mathematics classes benefit from communicating their thinking, using mathematical language. Students who are developing proficiency with mathematical language need opportunities to listen, speak, read, and write.

Teachers are encouraged to draw upon what students already know conceptually and should not assume that students understand mathematical terms. It is important for teachers to use new terms frequently in the context of class instruction and discussion. Students should be encouraged to define mathematical terms in their own words. By using symbols that are common to basic mathematics operations, teachers help ESL students ease into English.

Here are some techniques for working effectively with ESL/LEP students in *Basic Algebra*.

1. Use manipulatives such as place-value blocks and algebra tiles to model mathematical concepts. These materials can be linked to terminology and the students' own words and experiences. A hundreds board is useful for finding number patterns and extending counting skills.

2. Use graph paper and colored pencils to represent concepts such as those of fractions, decimals, and percents.

3. Students can create and label visual displays, charts, or graphs. Encourage them to illustrate symbols, factor trees, fractions, properties of numbers, decimals, and variables on index cards with corresponding definitions on the reverse side.

4. Create opportunities to respond to and use examples of mathematics from other cultures. Encourage all students to make up their own problems, enabling them to cast their stories in familiar cultural perspectives.

5. Build reinforcement with repetition, paraphrasing, and the use of synonyms. Vary discussion techniques by including short questions and answers that provide a change of pace and reduce the pressure of more complex sentence construction. During question-and-answer times, allow students extra time to respond so they can decode the English words and perform computations.

6. Take time to introduce words that have more than one meaning and might be confusing to students, such as *square*—a polygon with four equal sides and four 90-degree angles, and multiplying a number by itself.

7. Plan cooperative/collaborative activities, using structured techniques that combine small-group interactions with individual accountability. Sometimes pairing an ESL student with an English-proficient speaker in a peer tutoring interaction increases opportunities for understanding for both.

8. Some students may be learning the system of weights and measures for volume, length, time, temperature, and money for the first time. Hands-on practice helps them to overcome some of the difficulties.

9. Monitor student progress by reviewing work frequently and adjusting instruction to promote student understanding.

10. To give students practice reading application problems, use group activities in which students circle the important information or underline the words for mathematics operations.

11. Provide bilingual dictionaries, glossaries, and visual materials for classroom reference.

Cooperative Learning

Cooperative learning activities provide an alternative to traditional whole-class instruction and individual paper-and-pencil activities. Past practice emphasized individual computation and right answers. Systematic use of small groups in a cooperative learning atmosphere encourages a community of active learners working together. Students can help each other learn the material and extend their mathematical knowledge.

Small groups provide structures for students to exchange ideas, ask questions, and clarify concepts. Since mathematics problems can often be solved by more than one method, students can express different problem-solving strategies. As students explain their reasoning, they develop their use of mathematical language.

Cooperative learning is more than putting students together in small groups and giving them a mathematical task to solve. It involves careful attention to group process. The teacher can take an active role in structuring the learning situation by

- placing students in random or heterogeneous groups.
- providing structures for team building to encourage student collaboration.
- modeling appropriate social skills such as learning how to give encouragement and constructive feedback.
- rearranging the classroom so that groups have their own workspace.
- outlining the problem and learning objectives for students; letting them know what they are going to do, why they are doing it, and how they are expected to work.
- providing appropriate materials.

There are different ways to structure cooperative learning groups. Some teachers start mathematics class with students in groups to check homework. The expectation is that each student will have attempted each problem. Students are expected to deal with each other's questions and reach consensus on solutions.

Other teachers use small groups as a follow-up to whole-group instruction. Each group works on problems related to the lesson. Some cooperative learning strategies suggest that teachers assign roles to group members. There are a number of leadership and management functions such as recording information and tying ideas together by presenting conclusions.

Lessons that work best for cooperative learning have many possible answers or solution strategies. During cooperative learning time, the teacher circulates from group to group observing student interaction, offering assistance where needed, and asking questions to keep groups working productively. Individual accountability is fostered by establishing a pattern requiring that each student attempt to solve a problem. Students compare results and resolve differences. As teachers circulate, they can keep in touch with how students are responding to the mathematics content.

Time for summarizing is an important part of cooperative learning. As students share solutions and questions, the discussion can help them to generalize from a specific problem by looking for patterns or relationships in the data.

Newcomers to cooperative learning may find it helpful to work with a teacher who has been using the technique. Some school districts have formed support groups for sharing questions or problems.

Balanced approaches—whole group, small group, and individual—offer a variety of situations for students. A well-planned cooperative learning situation engages students in *actively* doing mathematics.

Manipulatives

The importance of manipulatives has been documented for students at the elementary level, but manipulatives have not been widely used to teach mathematics at the secondary level. Mathematics at the high school level moves to abstract and symbolic thinking, beyond the concrete thinking represented by manipulatives. Developmentally, high school students are expected to have reached a level of abstract and symbolic thinking beyond the need for concrete experiences.

Recently, the standards set forth by the National Council of Teachers of Mathematics have recognized the need for actively involving students in the development and application of mathematical concepts. They recognize that some students may need opportunities for informal activities and the use of concrete materials as they work toward a greater level of abstraction. Teachers can structure students' work with manipulatives toward that end.

Manipulatives provide multisensory learning experiences that allow students to model concepts and to move beyond specific examples to generalizations. The following are some tips to guide your use of manipulatives.

1. Choose materials that are appropriate for the concept you want to teach. No manipulative is adaptable to every situation.
2. Allow students time for independent exploration of a new material and plan activities that allow students to focus on the concept at hand.
3. As much as possible, provide materials for students to use rather than rely on teacher demonstration.
4. Use manipulatives in small group or paired working situations. Students learn from others as they work together.
5. Encourage students to verbalize and discuss ideas when they work with models.
6. Provide more than one material that can demonstrate a concept. As students experience concepts in more than one way, they move toward generalization and abstraction.
7. Model the use of appropriate mathematical language as you discuss students' ideas and questions.
8. As students explore a concept, help them record the observations they make while using manipulatives.
9. Remember that manipulatives do not teach by themselves. Materials, other students, carefully selected tasks, and teacher guidance are all important parts of the learning process.

Some manipulatives that work particularly well with *Basic Algebra* are:

- **Algebra tiles** The sizes and shapes of these pieces allow students to model equations, exponents, and integers.
- **Number lines and thermometers** Students can use these manipulatives to understand negative and positive numbers.
- **Calculators** Students should have time to explore calculators, to learn to use them as a tool, and to use them to reduce time spent on lengthy calculations with very large or small numbers.
- **Place-value blocks** These materials encourage students to construct models of place value, including whole numbers, decimals, and percents. Results can be recorded on grid or graph paper.
- **Fraction rectangles** Students who need further conceptual work with fractions will find that these manipulatives are good for modeling equivalent fractions and mixed numbers.

Problem Solving

Recent reports in mathematics research have suggested problem solving as the principal reason for studying mathematics. Reports from the workplace stress the need for workers who can use their mathematics skills and work collaboratively to solve problems. Our students need opportunities to develop problem-solving strengths.

We recognize the need to develop students' abilities to use a variety of strategies and techniques for solving problems. It is important to stress the *process* of problem solving. Students might learn how to solve particular problems when solutions are stressed. They are *more likely* to learn how to approach and solve other problems when process skills are also emphasized.

Problem solving usually includes these steps.

- Understanding the question.
- Finding the needed information.
- Planning what to do.
- Carrying out the plan.
- Checking the answers.

This model can help students see that problem solving requires several actions. Small groups can provide the setting for students rephrasing problems in their own words and clarifying their understanding of the question. Similarly, they can work through the other steps with ample opportunities to share their ideas and revise or expand their thinking.

This model provides an overview of steps involved in problem solving, but it does not suggest strategies or techniques for carrying out the steps. Here are some generalizations for problem-solving strategies.

- Problem-solving strategies can be taught. They help students explore possible solutions.
- No one strategy fits all problem situations. Some problems require more than one strategy.
- Students should have a number of strategies from which to choose. Encourage students to solve different problems with the same strategy, and the same problem with different strategies.

Some strategies for solving problems are:

- **Act it out.** This strategy helps students visualize what is involved in the problem. They go through the actions using manipulatives (or themselves) to make clearer the relationships among the parts of the problem.

- **Make a drawing or diagram.** This strategy provides a way to depict the information to make relationships apparent. Stress with students that elaborate drawings are not necessary. They should draw only what is essential to tell about the problem.

- **Look for a pattern.** The focus is on relationships between elements. Students may make a table and use it to find a pattern.

- **Construct a table.** This strategy can be useful for organizing data that helps students find a pattern or missing information.

- **Guess and check.** Students may have been discouraged from guessing in the past, but this can be a valuable strategy. Encourage students to make knowledgeable guesses and emphasize the need to check the answers. Reassure students that many guesses may be necessary and remind them that there may be more than one solution.

- **Work backwards.** Sometimes problems state the final conditions and ask about something that took place earlier.

- **Identify wanted, given, and needed information.** This stategy is useful in understanding the problem and the planning stages of the problem-solving model.

The process of checking answers is important for developing strong problem-solving skills. During this step, students look back at how they solved the problem, giving them the opportunity to discuss, clarify, or revise their thinking. Students generalize from one situation to another. Teachers can build on this process by asking students to find another way to solve the problem or find another solution.

Together, these approaches help students gain confidence in their ability to become mathematical problem solvers.

Chapter 1 The Language of Algebra

Introduction
Students may be surprised by the number of Americans who speak a particular language other than English. If there are any non-English–speaking or foreign-language students in your class, ask for examples of words in their language and compare them to the English translations. Make a list of several words in as many languages as possible. Then ask for examples of rules that explain how to put words together in sentences. Would the language be less effective as a tool of communication if such rules did not exist? How?

Lesson 1•1 Understanding Variables and Expressions

Chalkboard Activity
To assess students' prior knowledge, provide the following exercises on the chalkboard.

1. Find the variable in each variable expression.

   a. $2m + 15$ [m]   **b.** $\frac{w}{8}$ [w]

2. Evaluate these variable expressions when $x = 6$.

   a. $20 - x$ [14]   **b.** $\frac{24}{x}$ [4]

Working with ESL/LEP Students
The phrase *find the answer* may be more understandable than *evaluate the expression* to some students. Be sure, however, that students understand the meaning of an expression. Put it in terms of expressing, or communicating, an idea. Ask students, *What idea is being communicated by the expression 15n?* [There is something being multiplied by fifteen.]

Error Analysis
Students must understand that a variable can have different values. Use real-life examples to help them. For example, if two friends each have a different amount of money m, and each spends 75 cents, then the same variable expression, $m - 0.75$, can apply to both situations.

Final Check
Donna earns $6 per hour. Donna's boss deducts $9 for taxes each week. If she works h hours per day, four days per week, write an expression for her salary s. [$s = 4 \cdot 6h - 9$]

Lesson 1•2 Understanding Exponents

Chalkboard Activity
Challenge students to find as many ways as they can to express 32 with exponents. [$4 \cdot 8 = 2^2 \cdot 2^3$; $16 \cdot 2 = 4^2 \cdot 2$ or $2^4 \cdot 2$; $32 \cdot 1 = 2^5 \cdot 1$] Leave this exercise on the board throughout the lesson for students to attempt when they feel ready.

Using a Calculator
Use a calculator to evaluate 2^5.

2^5 [2 [∧] 5 [=] 32]

Allow students the opportunity to discover that the same answer is found three different ways.

For example, $2^5 =$

2 [x] 2 [x] 2 [x] 2 [x] 2 [x]

or

2 [x] 2 [=] [=] [=] [=]

or

2 [x] 2 [=] [x] 2 [=] [x] 2 [=] [x] 2 [=]

All three methods provide the same answer, 32.

Error Analysis
Make sure students realize that an exponent tells how many times to use a factor in a product. They should understand the difference between $3^2 = 3 \cdot 3$ and $2^3 = 2 \cdot 2 \cdot 2$.

Final Check
See if students can identify the base and the exponent in each expression.

1. x^2 [x, 2]
2. m^4 [m, 4]
3. 4^2 [4, 2]

Lesson 1•3 Using Grouping Symbols

Chalkboard Activity
Help students grasp the meaning of the grouping symbol *vinculum* introduced in the lesson. Use the following example to show why you find the value within the grouping first, as with parentheses. Bob cut lawns in the neighborhood, working 7 hours Saturday and 5 hours Sunday. He earned $60 for his hard work. How much did he earn per hour?

[$\frac{60}{7+5} = \frac{60}{12} = \$5$/hour]

Alternate Teaching Approach

Let students demonstrate for themselves the importance of doing the work above and below the vinculum before dividing. Suggest students evaluate the expression $\frac{15+9}{3+3}$ by performing the two division operations first, then adding. [5 + 3 = 8] Then evaluate the expression correctly by performing the addition, then dividing. [$\frac{24}{6}$ = 4] Obviously, the two values are not the same, and therefore, the order in which operations are performed matters.

Lesson 1•4   Understanding Order of Operations

Chalkboard Activity

Write 4 + 2 • 6 − 5 on the board and challenge students to find the answer without using calculators. Ask for answers; contribute one or two of your own if needed. After all possible results are given [6, 8, 11, 31, 43], ask how many students had each result. Invite one student from each group to show at the board how that particular result was obtained. These different results for the same problem will dramatize the need for rules for order of operations. [The correct answer is 11.]

Alternate Teaching Approach

Explain that the order of mathematical operations is strictly defined to avoid confusion. A mnemonic device for order of operations is **P**lease **E**xcuse **M**y **D**ear **A**unt **S**ally (parentheses, exponents, multiplication and division from left to right, and lastly, addition and subtraction from left to right).

Working with ESL/LEP Students

Explain the idea of a mnemonic device and help students create alternatives in their native languages.

Lesson 1•5   Using Inverse Operations

Chalkboard Activity

Invite students to make a list of actions that undo each other, such as going to school and going home by the same route; putting on shoes and taking them off; and borrowing money and paying it back. These actions are inverse, just as addition and subtraction or multiplication and division are, because one action undoes the other.

Using Manipulatives

Using algebra tiles, have students model adding 5 to x and subtracting 5 from (5 + x), multiplying x by 5 and dividing $5x$ by 5. Students could use the tiles to model exercises 1 to 4 in the lesson.

Lesson 1•6   Verbal Expressions and Algebraic Expressions

Chalkboard Activity

As you recite the following phrases, students could write them, using mathematical symbols.

1. Three more than half as many
   [$3 + \frac{1}{2}x$]

2. The sum of two numbers multiplied by four
   [$(x + y)4$]

3. Five feet more than twice the width
   [$2w + 5$]

4. Willy weighs fourteen pounds less than what Waldo weighs
   [$p - 14$]

Using Manipulatives

Using math tiles, have students show what $5x$ would look like (five 1-by-x tiles) and what $5 + x$ would look like (one 1-by-x tile and five 1s). Students could work in groups to model expressions with the tiles and have other students translate the expressions.

Working with ESL/LEP Students

Focus students' attention on the key phrases in the Word Expression chart in the lesson. Examples include *more than, the sum of, increased by,* and *the product of*. Spend extra time on the meaning of these phrases. Using props to act out the sample exercise about Jeff buying paint will help students translate between verbal and algebraic expressions.

Error Analysis

For subtraction and division, order matters, since the operations are not commutative. Students cannot always write down the problem as read or heard. The phrase "four less than a number" does not have the number 4 written first, as students write $4 - n$. This is an extremely common error. For clarity, one could say, "a number minus four." This is written exactly as stated.

Name _____ Date _____

Chapter 1   Test

Evaluate each expression for $a = 5$ and $b = 6$.

1. $a + 19$ _____
2. $8 \cdot b$ _____
3. $2 \cdot (b + a)$ _____
4. b^2 _____
5. $4b - a$ _____
6. $b^2 - a^2$ _____
7. Write $m \cdot m \cdot m \cdot m \cdot m$ using a base and an exponent. _____

Evaluate each expression.

8. 2^6 _____
9. 16^2 _____
10. $40 \div (12 \div 3)$ _____
11. $r \cdot (5 + r)$ for $r = 7$ _____
12. $34 - 36 \div 6$ _____
13. $18 - 9 + 4$ _____
14. $11 \cdot (5 - 2)$ _____
15. $16 + (4 + 3)^2$ _____

Name the operation that is the inverse of the one shown.

16. $21 + 5 = 26$ _____
17. 11 multiplied by a number _____
18. 45 subtracted from a number _____

Write an expression.

19. Let w represent the width of a rectangle. Write an expression to represent that the length is "5 feet longer than twice the width." _____

20. Chuck earns d dollars per hour. Write the expression that shows the amount of money he earns if he works 3 hours on Monday, Wednesday, and Friday, and 8 hours on Saturday.

_____

Chapter 2 The Rules of Algebra

Introduction
Discuss situations where rules are important. Answers might include automobile driving, sports games, or school rules. What happens when there are no rules? Would you want to drive if there were no rules? Could two teams compete in a sport if there were no rules?

Lesson 2•1   Recognizing Like Terms and Unlike Terms

Chalkboard Activity
Set up the following table on the board for students to complete. You might want to fill in the first row as a model.

	Number of Units	Like Units
5 ft + 3 yd − 5 in.	3	inches
4 quarters + 2 dimes	[2]	[cents]
5 km + 43 m + 0.8 km	[2]	[meters]
18 apples + 3 oranges	[2]	[fruit]

Using Manipulatives
Work in small groups. Use attribute blocks, pattern tiles, or a deck of cards. Students sort the items in at least two different ways. They write a description of the criteria they used to sort the items and the number of items in each group. Students can share their criteria and discuss the advantages and disadvantages of each system.

Working with ESL/LEP Students
Monetary examples may help students relate the meaning of like and unlike terms to their lives. If a passenger has five quarters and four dimes, they can use a like term, *cents*, to decide whether there is enough money to buy a train ticket for $1.50. Students can write an equation:

$$5 \times \$.25 + 4 \times \$.10 = \$1.25 + \$.40 = \$1.65$$

Students can practice converting different amounts of dimes and quarters to cents by writing similar equations.

Error Analysis
Students may add $2xy + 3xy$ and get an answer of $5x5y$ or $10xy$. Remind students that the coefficient tells how many of a particular article there are and that the rest of the term is the article. For example, $2xy$ means there are two xy's, not two x's and two y's.

Lesson 2•2   Understanding the Properties of Numbers

Chalkboard Activity
Write the terms $4xy$, $19x^2y$, $3x$, $5y$, $12xy$, $3x^2y^2$, $8x$, $5xy$ on the board. Then ask these questions:

1. How many terms are given? [8]
2. Name two like terms. [$4xy$ and $12xy$ or $3x$ and $8x$]
3. Name two unlike terms. [$4xy$ and $3x$ or $3x^2y^2$ and $5xy$]
4. What is the coefficient of the x^2y^2 term? [3]
5. What is the coefficient of the y term? [5]

Using Manipulatives
Manipulatives may help students explore abstract concepts that are difficult to grasp from verbal explanations. Encourage students to use algebra tiles to model the associative and commutative properties. For example, they can model $3 + 4$ with three tiles and four tiles, rearrange the tiles as $4 + 3$, and notice that the total stays the same.

Alternate Teaching Approach
As you discuss the properties, remind students that the names of the properties are words with which they may be familiar: *commute* describes traveling back and forth to work or school, or changing places; *associate* suggests something we do with friends, such as form groups; and *identity* describes oneself and is similar to the word *identify*.

Error Analysis
Students may have difficulty with the Multiplicative Property of Zero, confusing it with the Identity Property of Addition. Help students see that $5 + 0$ means adding zero of something to five of something, while $5 \cdot 0$ means having zero groups of five. Multiplying by zero changes the original number, while adding zero does not.

4   Chapter 2 *The Rules of Algebra*

Lesson 2•3 Understanding the Distributive Property of Multiplication over Addition

Chalkboard Activity
As a review, ask students to identify the properties below.

1. $4(3 + 5) = 4(5 + 3)$

    [Commutative Property of Addition]

2. $4(3 + 5) = 4 \cdot 3 + 4 \cdot 5$

    [Distributive Property]

3. $6(6 \cdot 7) = (6 \cdot 6) \cdot 7$

    [Associative Property of Multiplication]

4. Ask students to name one property that was not used above and give an example, using that property.

    [Possible answer: the Identity Property of Multiplication, $5 \cdot 1 = 5$]

Alternate Teaching Approach
Students working in small groups could use manipulatives to model the Distributive Property. They might arrange markers in one large array and then break it into two small arrays. Ask them to describe how the large array can be broken into the smaller ones. For example, an array of 36 can be 4 rows of 9, or 4 rows of 5 and 4 rows of 4.

Drawings can also show how the Distributive Property applies to multiplication of two numbers. For example: $25 \cdot 13$ could be drawn as:

	10	3
20	200	60
5	50	15

Adding the numbers in the boxes gives the sum of 325.

Error Analysis
In the problem $3(x + 2y)$ students may get the answer $3x + 2y$. Remind students to multiply the number outside the parentheses by each of the terms within the parentheses. An explanation of the meaning of *distribute* may help them. The number outside the parentheses is distributed among the terms inside the parentheses. An analogy to the distributor in a car (the part of the engine that distributes electricity to the spark plugs) may help some students grasp the meaning of the term in a mathematical sense.

Lesson 2•4 Adding Like Terms

Chalkboard Activity
Write this lattice multiplication on the board. Ask students if they can explain it and do another problem.

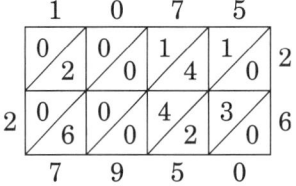

[This form of lattice multiplication shows that $1{,}075 \times 26 = 27{,}950$. It uses the Distributive Property by multiplying each place value separately and adding.]

Alternate Teaching Approach
The Distributive Property is the reason we can combine like terms and simplify algebraic expressions. Show students how to rewrite the following algebraic expressions, using algebraic properties.

1. $5x + 2x = (5 + 2)x$ Distributive Property
    $= 7x$

2. $m + 2n + 3m = m + 3m + 2n$
    Commutative Property
    $= (1 + 3)m + 2n$
    Distributive Property
    $= 4m + 2n$

Group some students together who are alike in some way. Ask students to figure out what is alike about the students in the group you chose. Then divide students into two teams. Each team creates a group of "like items" and challenges the other team to guess what they have in common.

Error Analysis
For students who have difficulty understanding why they can only simplify like terms, use the example of Harvey's commute time in the opening section of the lesson to illustrate what would happen if two different units of measurement were given equal value.

Lesson 2•5 Subtracting Like Terms

Chalkboard Activity
Evaluate each expression for $x = 1, 2, 3, 4,$ and 5.

1. $9x + 1 - 3x - 7 + x$
2. $7x - 6$

[Answers are the same for both expressions: 1, 8, 15, 22, and 29. Some students may realize the second expression is a simplified version of the first.]

Working with ESL/LEP Students
Work with students to help them understand the difference between a coefficient and an exponent. Use sample problems and make three columns: one each for exponents, variables, and coefficients. Then ask students to write the parts in the proper column.

Example: $3x^2 + 9y^4$

Exponents	Variables	Coefficients
2, 4	x and y	3 and 9

Error Analysis
If students simplify an algebraic expression such as $5r - r$ and get an answer of 5, they are forgetting that r in the expression has an understood coefficient of 1. Suggest that they interpret it as 5 items minus 1 item equals 4 items.

Final Check
Have students write two algebraic expressions that are equivalent, similar to the Chalkboard Activity. Then exchange expressions with a partner and evaluate them for values of x equal to 1, 2, 3, 4, and 5.

Lesson 2•6 Multiplying Terms

Using a Calculator
If students use a calculator to evaluate some algebraic expressions before and after they are simplified, it will help them to understand that they can apply the Associative and Commutative Properties to simplify expressions. For example, make a chart and evaluate each expression for the given values.

Evaluate. $3x^2(6x)$ and $18x^3$

If $x = 1$ $3 \cdot 1 \cdot 6 \cdot 1 = 18$ $18 \cdot 1 = 18$
If $x = 2$ $[3 \cdot 4 \cdot 6 \cdot 2 = 144]$ $[18 \cdot 8 = 144]$
If $x = 3$ $[3 \cdot 9 \cdot 6 \cdot 3 = 486]$ $[18 \cdot 27 = 486]$

Error Analysis
If students are having trouble identifying monomials, review the definition and explain that a product is the result of multiplying two or more numbers. Check to make sure they do not select expressions containing any other operations.

Final Check
Ask students to write a paragraph explaining the difference between adding or subtracting two monomials and multiplying two monomials. [Answers should emphasize that you can only add and subtract like terms, but you can multiply any two monomials.]

Lesson 2•7 Dividing Terms

Chalkboard Activity
Simplify and combine like terms.

1. $5(2m - 6n) + 4m$ $[14m - 30n]$
2. $7x(3x) - 4x(7y)$ $[21x^2 - 28xy]$
3. $5g(gh - h)$ $[5g^2h - 5gh]$
4. $a(10a) - 2a(5a)$ $[0]$

Alternate Teaching Approach
Challenge students to use multiplication to find the missing factors when they divide algebraic expressions. For example, $45x^2 \div 9x$. They can write a multiplication statement for this expression:
$? \cdot 9x = 45x^2$ and then think, *What factors do I need to have a product of $45x^2$?* The missing factors are 5 and x.

Error Analysis
To make sure students understand that the algorithms for multiplication and division apply to monomials only, ask them to solve these problems:

1. $(12 + 21) \div 3$ $[(4 + 7) = 11]$
2. $(x^2 + 6x) \div x$ $[(x + 6)]$

Both of these problems require students to use the Distributive Property and divide both terms within the parentheses by the divisor. It may be easier for students to understand the first example with numbers rather than the second example with variables, but emphasize that the process is the same for both examples.

Final Check
Challenge students to write two monomials, then find and label the sum, the difference, the product, and the quotient.

6   Chapter 2 *The Rules of Algebra*

Name _____ Date _____

Chapter 2   Test

Use this expression: $5y^3 + 4x + 3nq$

1. Name the terms in the expression. _____

2. Name the coefficients in the expression. _____

Simplify.

3. $4m - m$ _____

4. $17r - 4r$ _____

5. $y + y + 7y$ _____

6. $11cg - 9cg$ _____

7. $20k - (8k - 3k)$ _____

8. $a + a$ _____

9. $4x - 7y + 8x + 2y - 15$ _____

10. $3x^2 + 2y^2 - 5xy + x^2 - 8xy + 9y^2$ _____

Simplify each expression by multiplying or dividing monomials.

11. $4(6b)$ _____

12. $5x(4y)$ _____

13. $3a \cdot 7a$ _____

14. $(x)(9xy)$ _____

15. $24y \div 6$ _____

16. $\frac{17t}{t}$ _____

17. $\frac{13xy}{13x}$ _____

18. $\frac{36a^2bc}{9ab}$ _____

19. Find the sum, difference, product, and quotient of the monomials $24z$ and $6z$. _____

Name the property shown.

20. $w + 6v = 6v + w$ _____

21. $(x \cdot y) \cdot z = x \cdot (y \cdot z)$ _____

22. $t = t(1)$ _____

Use the Distributive Property of Multiplication Over Addition to write each expression without parentheses.

23. $6(2a + 3b)$ _____

24. $3x(x + 2y)$ _____

25. A rectangular table top measures $2n$ inches by $5n$ inches.
    a. Area = $l \cdot w$; find the area of the table top. _____
    b. Perimeter = $2l + 2w$; find the perimeter of the table top. _____

Chapter 3 Equations and Formulas

Introduction
Encourage students to think of other examples of balance in addition to that discussed on page 35. Examples include a gymnast on a balance beam, a game of tug-of-war, and the balancing of a food tray. For each example, ask students to describe what might cause an imbalance to occur. Students should recognize that an imbalance occurs when something is added to or removed from a balanced system. Conclude the discussion by challenging students to speculate how numbers might be balanced.

Lesson 3•1   What Is an Equation?

Chalkboard Activity
As a review of variables, ask students to simplify the following expressions. Save students' answers for Final Check.

1. $a + 5a + 3 + 2a + 1$ [$8a + 4$]
2. $4x + 4 - 3x - 1$ [$x + 3$]
3. $(3s)(6)$ [$18s$]
4. $\frac{12a^2}{4a}$ [$3a$]

Working with ESL/LEP Students
Before beginning the lesson, discuss the terms *true* and *false*. Have students identify in their native language the word for each. Encourage them to identify statements that are true and those that are false. You might also discuss how the meaning of "open" in the term *open sentence* differs from the meaning of "open" in *open door*.

Alternate Teaching Approach
Write numerals, operation symbols, variables, and equal signs on separate index cards. Allow students to arrange the cards in various ways to show true equations, false equations, and open sentences. Challenge students to change open sentences to true equations by solving for the variables.

Final Check
Refer to students' answers in the Chalkboard Activity. Ask students to change each expression to an open sentence that is equal to 36. Encourage them to solve for each variable.

1. $8a + 4 = 36$ [$a = 4$]
2. $x + 3 = 36$ [$x = 33$]
3. $18s = 36$ [$s = 2$]
4. $3a = 36$ [$a = 12$]

Lesson 3•2   Subtracting to Solve Equations

Chalkboard Activity
Ask students to label each equation as true, false, or open. Encourage them to use mental math to solve each open equation.

1. $8 + 8 = 17$ [false]
2. $r + 100 = 108$ [open; $r = 8$]
3. $s + 19 = 30$ [open; $s = 11$]
4. $45 + 24 = 70$ [false]
5. $n + 2.5 = 7$ [open; $n = 4.5$]
6. $5.5 + 6 = 11.5$ [true]

Working with ESL/LEP Students
Discuss the meaning of *inverse* with students. Suggest that students make a two-column chart in which they list each of the four operations—addition, subtraction, multiplication, and division—in one column and its inverse operation in the other column. Students can use symbols instead of words for the operations. Encourage them to save their charts to refer to as needed.

Alternate Teaching Approach
Algebra tiles can be used to model open sentences in which students subtract the same number from both sides. The drawing below models the equation $8 + x = 13$. By subtracting 8 small tiles from each side of the equation, students will observe that $x = 5$.

```
□□□□□□□□ + ▢ = □□□□□□□□□□□□□
    8       + x =        13
```

Error Analysis
In exercise 9, some students may have difficulty subtracting 4.5 from 8. You may wish to review place value and decimals with those students.

Final Check
To help students understand why subtraction is used to solve equations containing addition, discuss the following questions.

1. What is the inverse operation for addition? [subtraction]

8   Chapter 3 *Equations and Formulas*

2. How do inverse operations help to solve equations? [Possible answer: inverse operations help to isolate the variable.]

Lesson 3•3 Adding to Solve Equations

Chalkboard Activity
As a review, ask students to solve the following equations of the form $x + a = b$. Before beginning, you might ask them to identify the inverse operation they will use to solve each equation. [subtraction]

1. $x + 11 = 15$ [$x = 4$]
2. $z + 4.2 = 7.3$ [$z = 3.1$]
3. $s + 21 = 43$ [$s = 22$]
4. $21 = e + 8$ [$e = 13$]

Using Manipulatives
Use counters to explain how addition is used to solve some equations. Encourage students to use the counters to set up their own problems and solve them.

Alternate Teaching Approach
Students can work in pairs. Give each pair several index cards with equations of the form $x - a = b$ written on them. One student solves the equation, and the other student checks it. Students then repeat the activity, reversing roles. When all equations have been solved, challenge students to make up their own equations and exchange them with their partner.

Final Check
Invite students to write two equations: one that requires addition to solve and another that requires subtraction. Encourage students to solve the equations and then check them.

Lesson 3•4 Dividing to Solve Equations

Chalkboard Activity
As a lead-in to this lesson, encourage students to use mental math to solve each equation. Save students' work on the chalkboard to refer to in Final Check.

1. $5x = 15$ [$x = 3$] 2. $\frac{1}{2}r = 2$ [$r = 4$]
3. $9s = 81$ [$s = 9$] 4. $120 = 6a$ [$a = 20$]
5. $1.5k = 3$ [$k = 2$] 6. $8t = 320$ [$t = 40$]

Ask students whether they could use mental math to solve the equation $3.45x = 14.49$. [Most students will probably say "no."] Tell students that this lesson will show them how to solve multiplication equations that are difficult to solve mentally.

Alternate Teaching Approach
Write the following equations on the chalkboard to show that multiplying by the multiplicative inverse of a number is the same as dividing by the original number.

Multiplying by multiplicative inverse:

$$\frac{3}{4}x = 12$$
$$\frac{4}{3} \cdot \frac{3}{4}x = \frac{4}{3} \cdot 12$$
$$1x = \frac{4}{3} \cdot 12$$
$$x = \frac{48}{3}$$
$$x = 16$$

Dividing by original number:

$$\frac{3}{4}x = 12$$
$$\frac{3}{4}x \div \frac{3}{4} = 12 \div \frac{3}{4}$$
$$\frac{3}{4}x \cdot \frac{4}{3} = 12 \cdot \frac{4}{3}$$
$$1x = \frac{48}{3}$$
$$x = 16$$

Final Check
Return to the equations in the Chalkboard Activity. This time, ask students to solve the equations by using division. Encourage them to show all of their work.

Lesson 3•5 Multiplying to Solve Equations

Chalkboard Activity
Write the open sentences $4x = 12$ and $\frac{x}{4} = 12$ on the chalkboard. Ask for a volunteer to read each sentence. Be sure the second sentence is read "x divided by 4," not "x over 4." Ask another volunteer to explain what inverse operation should be performed on the first equation to find a solution. Invite others to guess what inverse operation should be performed to solve the second equation. Lead students to recognize that multiplication can be used to solve certain equations. You might repeat this activity with two other equations, such as $3c = 33$ and $\frac{c}{3} = 33$.

Using Manipulatives
Use counters to explain how multiplication can be used to solve some equations. Students can

Chapter 3 *Equations and Formulas* 9

describe the steps as you demonstrate how to solve the equations. Encourage students to use counters to solve their own equations.

Using a Calculator
Students can use calculators to solve exercises 20 to 24. You might help them solve one of the exercises, for example, $\frac{m}{25} = \frac{9}{15}$. Explain that they can do the calculation several ways. First, they can multiply 9 and 25, then divide by 15. Second, divide 9 by 15, then multiply by 25. Third, divide 25 by 15, then multiply by 9. Each method gives 15 as the answer. Encourage students to verify that the three calculations lead to the same answer.

Error Analysis
For exercise 26, students who choose to multiply both sides of the equation and then subtract may not realize that both $\frac{x}{2}$ and 4 must be multiplied by 2, not just $\frac{x}{2}$. You may wish to take the time at this point to review the Distributive Property of Multiplication Over Addition.

Final Check
Suggest that students repeat exercises 20 to 24, using the cross-products method. Encourage them to compare their answers with those they obtained earlier by multiplying both sides of the equation by the same number.

Lesson 3•6 What Is a Formula?

Chalkboard Activity
As a lead-in to the lesson, ask students how far a biker will travel in 3 hours if she is traveling at a speed of 15 miles per hour. [45 miles] Then on the chalkboard, write other rates of speed and traveling times for the biker. Ask students to use the information provided to calculate the distance traveled by the biker in each case. Encourage students to come up with an open sentence that describes the relationship between the biker's rate of speed, time traveled, and distance traveled. At this time accept any letters for variables. Tell students that the equation they wrote is called a formula and that they will learn more about formulas in the lesson.

Working with ESL/LEP Students
Discuss the meaning of *formula* with students. Encourage them to relate in their native language as well as in English several examples of formulas. Review the concepts of area, perimeter, rate, distance, discount, and sale price and the symbols used to represent each.

Alternate Teaching Approach
Before doing exercises 8 to 11, you might have students read the formulas aloud, substituting words for symbols. For example, for exercise 8, students might say, "Distance equals rate of speed times time. If the distance traveled is 45 and the time traveled is 5, what is the rate of speed?"

Error Analysis
Some students confuse the formula for finding the area of a rectangle with the formula for finding its perimeter. Point out that the area formula involves multiplying two lengths, and the answer is in square units. The perimeter formula involves adding lengths, and the answer is not in square units but in linear units. To reinforce the difference between the two formulas, you might have students solve perimeter and area problems that involve squares and triangles.

Lesson 3•7 Focusing on Equations with Fractions and Mixed Numbers

Chalkboard Activity
Write the following equations on the board.

1. $x - 7 = 5$
2. $\frac{3}{4}x - 7 = 5$
3. $2\frac{1}{2}x - 7 = 5$

Challenge students to describe how the equations differ. [The coefficient of *x* is 1 in equation 1, a fraction in equation 2, and a mixed number in equation 3.] Ask what the first step would be in the solution for all three exercises. [Add 7 to both sides.]

Using Manipulatives
Fraction bars can be used to review mixed numbers and improper fractions. Provide pairs of students with construction paper and a ruler. Encourage them to draw fraction bars to show that $\frac{3}{3} = 1$, $\frac{8}{3} = 2\frac{2}{3}$, and $3\frac{1}{3} = \frac{10}{3}$.

Error Analysis
In exercises 6 and 9, some students may add or subtract fractions with different denominators by adding or subtracting the denominators as well as the numerators. Remind students that to add such fractions, they must first find the fractions' common denominator. Students might benefit from a review of least common multiples and least common denominators at this point.

10 Chapter 3 *Equations and Formulas*

Name _____ Date _____

Chapter 3 Test

State whether each equation is *true*, *false*, or *open*.

1. $20 - 14 \div 2 = 13$ _____
2. $11 = x + 17$ _____

Which of the given values is the solution to the equation?

3. $6a = 30$; 5, 8, 24, 36 _____

Substitute the values of the variables into the formula. Then solve the equation for the remaining variable.

4. $A = lw$; $A = 35$, $l = 14$ _____
5. $s = r - d$; $s = 26$, $d = 9$ _____
6. $d = rt$; $d = 10.5$, $r = 1.5$ _____

Name the operation you can use to solve each equation.

7. $12f = 84$ _____
8. $b + 19 = 11$ _____

Give the multiplicative inverse of each number.

9. $\frac{1}{8}$ _____
10. $\frac{7}{5}$ _____
11. 10 _____
12. $1\frac{3}{4}$ _____

Solve each equation.

13. $7p = 84$
14. $h - 1\frac{1}{2} = 1\frac{1}{4}$
15. $17 = \frac{1}{4}w$

16. $t + 5.9 = 13.4$
17. $\frac{n}{12} = 10$
18. $\frac{5}{6}c = 204$

19. Lourdes borrowed $2,200 from her brother and has repaid $700. She intends to pay the balance in equal payments over the next 12 months. Write and solve an equation describing the situation.

    _____

20. In a 55-mile-per-hour speed zone, Tony traveled a distance of $37\frac{1}{2}$ miles in $\frac{3}{4}$ hour.

    a. Write the formula you can use to find Tony's average speed.

    _____

    b. Substitute the values of the variables into the formula.

    _____

    c. Solve the equation for Tony's average speed.

    _____

Chapter 4   Integers

Introduction
Temperature ranges vary across the United States. Discuss the high and low temperatures for your region. Then look at the motor oil chart and decide which one would be the best one to use in your region. What other products depend on the temperature range for effectiveness? [Possible answers include types of clothing, heating and cooling equipment, and some sports equipment.]

Lesson 4•1   What Is an Integer?

Chalkboard Activity
Draw part of a simple city map on the board, including several city blocks. Choose an intersection on the map as a starting point and count off a certain number of blocks in one direction. Ask students how you should mark off the same distance but in the opposite direction of the starting point. [Students should describe moving the same number of blocks, but in the opposite direction from the starting point.]

Choose another intersection and ask volunteers to mark off the same distance in opposite directions from the starting point. Then ask, *If you walk a certain distance from the intersection, how many places are the same distance but in the opposite direction of the intersection?* [one] Relate the movement along a street to movement along a number line.

Using Manipulatives
Algebra tiles are useful when discussing integers. Use one color to represent positive quantities and another color to represent negative quantities. As an introductory activity, use an overhead projector. Students can show, with tiles, the opposites of the numbers you show on the overhead. For example, if you put 3 red tiles on the overhead, they should show 3 blue tiles.

Working with ESL/LEP Students
"Simon Says Opposites" will help students understand the meaning of the word *opposite*. Players do the opposite of what Simon says. Begin with all players standing, then give simple directions such as *sit down, stand up, raise your right hand,* and *turn around*. Students need to think of and do the opposite activity. Let each student have a turn being leader.

Error Analysis
For exercises 12 to 14 some students may require discussion of what integers can represent. Remind them that an increase or decrease of 20, for example, could relate to many things such as dollars, degrees, stories in a high-rise building, or depth of a mine shaft.

Lesson 4•2   Integers and the Number Line

Chalkboard Activity
Write the opposite of the given integer.

1. 5 [−5]   2. positive 75 [negative 75]
3. −3 [3]   4. 200 [−200]
5. 500 [−500]   6. negative 10 [positive 10]
7. −35 [35]   8. −1,122 [1,122]

Alternate Teaching Approach
Use masking tape to make a number line on the floor. Write an integer on a slip of paper for each student. Students can take turns locating their integer on the number line and standing on it. After each student has had a chance to locate a number on the line, change the activity by letting a student choose a spot, then asking another student to stand on the opposite of that number.

Working with ESL/LEP Students
Some students may be confused between positive and negative. It might help to color code the number line by writing positive numbers in red and negative numbers in blue, with an enlarged *0* point.

Error Analysis
Some students may forget that the numbers to the right of zero represent positive numbers even if they do not usually have a plus sign. Numbers to the left of zero represent negative numbers and always have a negative sign.

Lesson 4•3   Absolute Value

Chalkboard Activity
Draw a number line on the chalkboard with only zero marked. Try the following verbally, or write on the board and ask students to select the correct answer.

Which is farther from zero?

1. −4 or 5? [5] 2. 3 or −8? [−8]

3. 10 or −2? [10] 4. −6 or 1? [−6]

Using Manipulatives
Use the number line from the Chalkboard Activity or make one with masking tape on the floor. Ask a student to find a point 4 units from zero. After the first student finds a spot, ask if anybody can find another point 4 units from zero. Continue until students begin to realize that there are always two points the same distance from zero. Then explain that the distance from zero is called the absolute value of a number.

Working with ESL/LEP Students
Some students may not understand why the absolute value cannot be a negative number. Review with students the meaning of absolute value. Distance is always a positive quantity.

Lesson 4•4 Which Integer Is Greater?

Chalkboard Activity
Display a number line. Write each expression on the board as a review of absolute value and in preparation for comparing integers.

1. $|6| + |-1|$ [7]

2. $|-13| - |-2|$ [11]

3. $4(|-2| + |4|)$ [24]

Alternate Teaching Approach
Students can work in pairs to play this game. Provide each group with a set of cards consisting of >, <, = signs, and the integers from −10 to 10. Students take turns turning over two integer cards and placing the correct symbol between them. Challenge students to turn over five integer cards and then arrange them in order from least to greatest.

Error Analysis
For exercises 14 to 19 students should use a number line. Tell them to find each number in the set on the number line. Remind them that the number that is farthest to the right is the greater number.

Lesson 4•5 Adding Integers

Chalkboard Activity
Write each set of integers on the board. Students can rewrite the set with the integers in order from least to greatest.

1. {4, −1, −3} [−3, −1, 4]

2. {7, −3, −2} [−3, −2, 7]

3. {1, −1, 2, −2} [−2, −1, 1, 2]

4. {7, 2, 5, −5} [−5, 2, 5, 7]

Using Manipulatives
Use algebra tiles to add integers. Use red (or another color) tiles to represent positive quantities, and blue (or another color) tiles to represent negative quantities. To add −4 + −2, students place 4 blue tiles and 2 blue tiles together. They can see that the sum is −6. To add −4 + 3 arrange 4 blue tiles above 3 red tiles. Whenever a blue and a red tile are together, they cancel each other, since they are opposites. Only one blue tile would remain, so the answer is −1.

Working with ESL/LEP Students
Provide students with one large number line. Demonstrate how to add integers by counting on the number line. Always start at zero. Practice moving to the right for positive and to the left for negative. Encourage students to count on the number line for each addition problem until they are confident with more abstractions.

Error Analysis
For some of the Practice exercises, students may have difficulty with the larger numbers because their number lines are not long enough to use. Encourage them to act out the process while using an imaginary number line. For example for 25 + −8, they can think of going 25 places to the right, and only going 8 places to the left. Students should realize that they are still on the right side of zero, but have moved backwards 8 places from 25, which is 17.

Lesson 4•6 Subtracting Integers

Chalkboard Activity
Write the following problems on the board, and challenge students to find the missing numbers. These exercises should help students understand the subtraction rules, since finding the missing addend is subtracting.

1. 11 + ___ = 18 [7]   2. 11 + ___ = 6 [−5]

3. 11 + ___ = 0 [−11] 4. 11 + ___ = −1 [−12]

5. −3 + ___ = −1 [2]   6. −3 + ___ = 1 [4]

7. −3 + ___ = 6 [9]   8. −3 + ___ = −6 [−3]

Using Manipulatives

Use algebra tiles to subtract integers. To find the difference: $6 - (-3)$, put 6 red tiles on the overhead. Ask students, *How can I subtract 3 blue tiles, if I only have red tiles?* If nobody suggests it, show them that you can add 3 red and 3 blue tiles to the overhead. Since the 3 red and 3 blue are opposites, they cancel each other, so you haven't changed the original value of positive 6. Now if you remove 3 blue tiles, there are 9 red tiles left. This shows that $6 - (-3) = 9$.

Alternate Teaching Approach

Make a number line on the floor. Have students take turns acting out subtraction problems. Each student should start at zero, facing the positive direction. To act out $-3 - 5$, say: "*Move 3 units backward* (the -3). *Now face the opposite direction* (the subtraction sign). *Move forward 5 units* (the 5)." The student should be standing on -8.

Error Analysis

For exercises 15 to 18, some students may need to be reminded that addition and subtraction are inverse operations. To solve the exercise $8 - \_\_\_ = 5$, students should think or write an addition statement such as: $? + 5 = 8$.

Lesson 4•7   Multiplying Integers

Chalkboard Activity

Students can draw a number line to show each addition or subtraction statement. The first one is shown.

1. $-7 - (-5) = -2$

    $-7 + 5 = -2$

    (number line from -7 to 7)

2. $-5 + -8 = -13$
3. $6 - 8 = -2$
4. $5 - (-2) = 7$

Using a Calculator

Most calculators have a [+/−] key for negative numbers. Students can experiment with their calculators to see how to enter negative numbers. Try entering 5 [+/−] [+] 3. Students should get -2. Sometimes the subtraction key will also function as a negative sign. Try the same problem, using the [−] key; it should give the correct answer. Then try reversing the problem using the [−] key, such as $5 - (-3)$. Students will probably get an incorrect answer of 2. Now try several multiplication problems, such as $-5 \cdot 3$. The answer should be -15 if they use either the [+/−] key or the [−] key. Then reverse the signs ($5 \cdot -3$). Using the [−] key will probably give 2 as an answer, but the [+/−] key will show the correct answer of -15.

Working with ESL/LEP Students

Have students write the rules for addition, subtraction, and multiplication of integers on index cards, with an example following each rule. Students can use the cards for reference when working on the exercises.

Final Check

Challenge students to find two integers whose sum is -10 and whose product is 21, similar to exercise 17 on page 73. Ask them to explain how they found their answers.

Lesson 4•8   Dividing Integers

Chalkboard Activity

Write the numbers below on the board. Challenge students to write addition, subtraction, and multiplication problems that have each number as an answer.

20, 25, -12, -18, and 42
[possible answers for 20:
10 + 10; 32 − 12; and 4 • 5]
[possible answers for -18:
$-9 + -9$; $-10 - 8$; and $-3 \cdot 6$]

Alternate Teaching Approach

Use a number line to help students visualize division with positive and negative numbers. Start with a basic fact such as $12 \div 4$. Mark off 4 groups of 3 units on the number line between 0 and 12 to show that $12 \div 4 = 3$. Then do the same thing with $-12 \div 4$ on the left side of the number line. There are 4 groups of -3 in -12. Use the number line to show $-12 \div -4$. There are 3 groups of -4 units in each group.

Error Analysis

Exercises 17 to 20 require students to remember the relationship between multiplication and division. Students might need to rewrite each division exercise as a multiplication problem. For example, to solve $-12 \div ? = -4$, they should write $? \cdot -4 = -12$. Writing the exercise in this format will be easier for most students.

14 Chapter 4 *Integers*

Chapter 4 Test

1. Write the opposite of −13. _____

Use the number line to complete exercises 2 and 3.

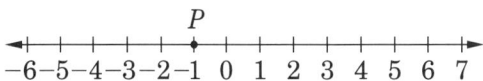

2. Write the number represented by point P. _____
3. Mark the point −5 on the number line and write the letter Q above it.

Find each absolute value. Simplify.

4. $|-7|$ _____
5. $|9|$ _____
6. $|-4| + |-12|$ _____
7. $|8| + |-9|$ _____
8. $|-3 + -14|$ _____
9. $|-6 + (-5)|$ _____

Write the numbers in order from least to greatest.

10. $\{1, -4, 0, 3, -1\}$ _____
11. $\{2, -2, 1, -4, 4\}$ _____

Find each sum, difference, product, or quotient.

12. $8 - (-3)$ _____
13. $(-5)(-6)$ _____
14. $(-6) + 2$ _____
15. $-21 \div 3$ _____
16. $18 + (-11)$ _____
17. $7(-4)$ _____
18. $(-18) \div (-2)$ _____
19. $(-22) - 17$ _____
20. $(-14)5$ _____
21. $(-7) - (-8)$ _____
22. $0 - (-4)$ _____
23. $-72 \div (-6)$ _____

Write >, <, or = to make a true equation or inequality.

24. $8 - 15$ _____ $(-6) - 1$
25. $9(-3)$ _____ $(-22) \div 2$

Chapter 5 Introduction to Graphing

Introduction

In order to graph coordinates successfully, students should be able to read data both horizontally and vertically. To this end, present additional questions that require the use of the nutritional values chart. For example, What food has 16 g of protein? *[1 cup of lentils]* How many calories are in a small peach? *[37] Students can make up their own questions, using the chart.*

Lesson 5•1 The Coordinate Plane

Chalkboard Activity
Write the following headings: "Parallel Lines" and "Perpendicular Lines." Ask students to provide several concrete examples of each term. Encourage them to look around the classroom for ideas.

Working with ESL/LEP Students
Students may have difficulty with directional terms such as *northwest*. You may want to provide a compass rose labeled in English and invite students to write the corresponding directional terms in their native language. Also use directional terms to describe the layout of the school grounds, e.g., *The parking lot is northwest of the classrooms.*

Alternate Teaching Approach
If your classroom has floor and/or ceiling tiles, students can discuss the grid formed by the borders between the tiles. Invite them to devise a system for naming points on the grid.

A game like "Battleship" could help students with the concept of coordinates. The game calls for one player to place a battleship at particular, undisclosed coordinates on their "ocean" grid. Then the opponent tries to sink the battleship by firing torpedoes at specific coordinates. The player announces "hit" or "miss." The first person to sink the opponent's ship wins the game.

Lesson 5•2 Ordered Pairs

Chalkboard Activity
Draw a map grid on the board, using letters and numbers for the axes. Then draw objects such as a river, park, and several buildings on the grid. Ask students to locate the objects as you call out coordinates. Then ask them to name the coordinates as you call out the objects.

Alternate Teaching Approach
If students' desks are in rows and columns (or can quickly and easily be arranged that way), have them describe where they are sitting by giving an ordered pair. The first number in the ordered pair will be the row number and the second number will be the column or seat number. Demonstrate that there is a difference between the student who sits in (3, 2) (row 3, seat 2) and the student who sits in (2, 3) (row 2, seat 3).

Error Analysis
To help students remember that the x-coordinate always comes first in an ordered pair, point out that x also comes before y in the alphabet.

Lesson 5•3 Tables and Graphs

Alternate Teaching Approach
Ask students to bring in take-out menus from restaurants or advertisements from the newspaper. Working in small groups, students should choose an item they would like to buy or order. Group members can then make a table that shows the cost of one item, two items, etc. (For example, one item costs $2.25 or two items cost $4.50 and so on.) Encourage students to use the information in their tables to make graphs. Point out that each possibility in the table can be expressed as a point, with the number of items being the x-coordinate and the total cost being the y-coordinate.

Error Analysis
Be sure students understand that the information in the third column of each table in exercises 5 and 6 comes from the first two columns. No additional computation is needed once they complete the second column.

Lesson 5•4 Special Graphs

Chalkboard Activity
Challenge students to draw four lines, without having to lift up their pencil, so that all nine dots are passed through.

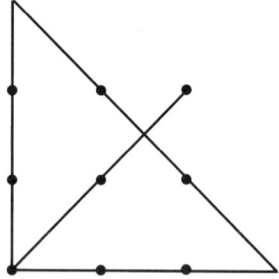

Alternate Teaching Approach

If you did not use the Alternate Teaching Approach for Lesson 5•3, you may want to try it now. If students have done the activity, ask them to refer to their tables and graphs. Ask whether the points they plotted should be connected. For example, if three sandwiches cost $3.75 and four sandwiches cost $5.00, connecting the point (3, 3.75) to (4, 5) would imply that 3.5 sandwiches would cost between $3.75 and $5.00. That may be true, but most restaurants will not sell 3.5 sandwiches. Therefore, the points should not be connected.

Error Analysis

Students may have trouble understanding the difference between a continuous and a discrete graph, even if they can state the definitions. To explore students' understanding, have them change the discrete graph for Crispy Rice to a continuous graph. Then ask for volunteers to explain what this change would mean in terms of buying cereal. [Any fractional part of boxes of cereal could be sold.]

Final Check

To assess students' understanding of the terms introduced in this lesson, discuss the following questions.

1. How are continuous graphs and discrete graphs similar? How are they different? [Both involve points plotted on a coordinate plane, but only continuous graphs have a line passing through the points.]

2. What do you know about all the points on the graph of a vertical line? [They have the same x-coordinate.]

Lesson 5•5   Slope

Chalkboard Activity

Draw several different right triangles to represent different slopes. Encourage students to discuss what walking, skateboarding, sledding, and mountain biking might be like on a hill with each slope.

Alternate Teaching Approach

Some students may have more success if this lesson is presented in terms of the slopes of hills rather than lines on a coordinate plane. The application of the different slopes can be presented in terms of the ease of climbing, as in the Final Check below.

Error Analysis

Students often invert the ratio that expresses the slope of a line. You may want to mention that another way of expressing the slope of a line is $\frac{\text{rise}}{\text{run}}$. Rise is the movement up or down along the y-axis, and run is the movement left or right along the x-axis.

Final Check

Present the following slopes: $\frac{1}{2}, \frac{1}{3}, \frac{1}{4}, \frac{1}{5}$. Tell students to suppose these slopes represent the steepnesses of four hills. Which hill would be easiest to climb? [hill with slope $\frac{1}{5}$] Challenge students to draw these slopes on a coordinate plane and to label each line with its slope.

Lesson 5•6   Finding the Intercepts of an Equation

Chalkboard Activity

As a review, ask students to find the slope of the line through the given points.

1. (5, 2) and (−3, 4) [$\frac{1}{4}$]
2. (−1, −3) and (−5, −2) [$\frac{1}{4}$]
3. (6, 7) and (4, 5) [1]
4. (7, 2) and (9, 8) [3]

Using a Calculator

Ask students to graph the equations $y = 3x + 5$, $y = -\frac{1}{2}x + 5$, $y = 4x + 5$, and $y = \frac{2}{3}x + 5$. As students describe the graphs, they should note that all the graphs pass through the point (0, 5), but slant differently. Then have the students graph the equations $y = 2x + 5$, $y = 2x - 5$, $y = 2x + 3$, and $y = 2x$. Students should note that all the graphs slant the same way, but cross the y-axis at different points.

Error Analysis

Students may have trouble finding the x- and y-intercepts for lines as given in exercises 8 and 9. When finding the y-intercept in exercise 8, students substitute 0 for x, but may write $4 - 3y = 12$, $-3y = 8$, $y = -\frac{8}{3}$. Emphasize that $4x$ means multiply, and that $4 \cdot 0 = 0$.

Name _____ Date _____

Chapter 5   Test

Name the coordinates of each point on the graph.

1. A _____   2. B _____

On the graph draw and label each point.

3. $C(3, 5)$ 4. $D(-4.1, -2.3)$

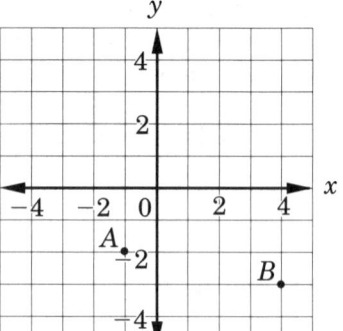

Name the quadrant in which each point is located.

5. $(-1, 4)$ _____   6. $(2, 3)$ _____

Find the distance between each pair of points.

7. $(3, -5)$ and $(3, 4)$ _____   8. $(12, -6)$ and $(2, -6)$ _____

Use the equation $y = -5x - 3$. Find the value y for each given value of x.

9. $x = -2$, $y = $ _____   10. $x = 4.3$, $y = $ _____

11. Complete the table of values for the equation $y = 2x - 5$. Then graph the equation.

x	y	(x, y)
3		
0		

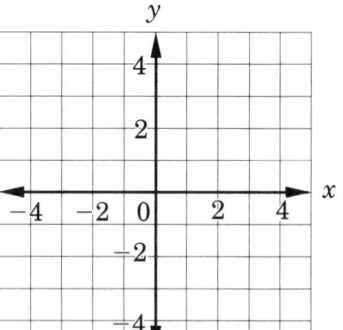

12. A vertical line passes through the point $(6, -3)$. What is the equation of the line? _____

13. Find the slope of the line passing through the points $(-3, -2)$ and $(4, 7)$. _____

Find the slope of the graph of each linear equation.

14. $y = \frac{2}{5}x + 4$ _____   15. $6x + 2y = 12$ _____

Find the x-intercept and the y-intercept of the graph of each equation.

16. $y = 3x + 5$ _____   _____

17. $4x - 2y = 8$ _____   _____

18. $8x + y = 16$ _____   _____

19. $6y = 12x$ _____   _____

20. $y = 8$ _____   _____

18   Chapter 5 Test

Chapter 6 More About Graphing

Introduction
After discussing the graphs on page 95, ask students for other specific objects or events that might be represented by graphs drawn on a coordinate plane. Answers might refer to part of a roller coaster ride or the path of the moon across the night sky.

Lesson 6•1 Slope as Rate of Change

Chalkboard Activity
To introduce the idea that the slope of a line is a rate of change, have students describe the line that passes through each pair of points.

1. (0, 0) and (2, 2) [slope is 1; passes through origin]
2. (−1, 5) and (−1, 2) [vertical line; no y-intercept]
3. (2, 5) and (−1, 5) [horizontal line; y-intercept is 5]
4. (3, −3) and (−3, 3) [slope is −1; passes through origin]

Working with ESL/LEP Students
Be sure students understand what the term *rate* means. Challenge them to make a list of phrases that contain the word *per*. [Possibilities include $5.00 per hour for wages, miles per hour, revolutions per minute, and 1 coupon per person.] Encourage students to work in small groups for this activity. Groups can share and compare their examples. If students have difficulty, then you may want to suggest that they look through newspapers and magazines for examples.

Alternate Teaching Approach
Emphasize the relationship between slope and rate of change. The slope of a line can be expressed as the rate of change between two points on the line. For example, if the slope is $\frac{3}{4}$, the rate of change between the points can be expressed as 3 vertical units for (or per) every 4 horizontal units.

Final Check
To assess students' understanding of slope and rate of change, discuss the following questions.

1. How is finding the slope of a line similar to finding the annual rate of change in a city's population? [Both involve comparing the change in one quantity to the change in another quantity.]

2. How is finding the slope of a line different from finding the annual rate of change in a city's population? [Students should realize that the process is the same, up to a point. Finding the rate of change in a city's population involves the extra step of dividing the ratio so that the denominator is 1 (a single unit).]

Lesson 6•2 Using the Slope-Intercept Form of an Equation

Chalkboard Activity
As a review, challenge students to find the rate of change (per year) in the price of each item. Ask volunteers to explain how they got their answers.

Item	Price Five Years Ago	Price Today
A	$1.20	$2.10
B	$59.95	$74.95
C	$7,827	$9,185

[item A: $0.18/y; item B: $3/y; item C: $271.60/y]

Alternate Teaching Approach
Graph the following equations on the chalkboard or on an overhead graphing calculator: $y = 3x$, $y = 3x + 1$, and $y = 3x + 4$. Ask students to study the graphs and to describe how they are similar and different [same slope but different y-intercepts]. Then graph these equations: $y = -x - 3$, $y = \frac{1}{2}x - 3$, and $y = -3x - 3$. Point out that the lines in these graphs have different slopes but the same y-intercept.

Error Analysis
A mistake students are likely to make in exercises 2 and 3 is forgetting to solve for y. Stress that an equation is only in slope-intercept form if the coefficient of y is 1.

Final Check
To assess students' understanding of the slope-intercept form of an equation, ask and discuss the following questions.

1. How would you use the slope-intercept form to graph $2x - 3y = 6$? [Students' answers should reflect their understanding that the equation needs to be rewritten in slope-intercept form: $y = \frac{2}{3}x - 2$.]

2. The y-intercept of a line is (0, 1). How would knowing that the slope of the line is $\frac{5}{8}$ help you find a second point on the line? [A slope of $\frac{5}{8}$ means you can move up 5 units and to the right 8 units from (0, 1) to find a second point.]

Lesson 6•3 Other Forms of Equations

Chalkboard Activity
Ask students to find the slope and y-intercept of each line.

1. $y = 4x - 2$ [4; −2]
2. $6y = 2x + 1$ [$\frac{1}{3}$; $\frac{1}{6}$]
3. $8x = 3y + 5$ [$2\frac{2}{3}$; $-1\frac{2}{3}$]
4. $y = -3x$ [−3; 0]

Alternate Teaching Approach
Students could benefit from seeing at least one nonmonetary example of a real-world situation that can be expressed with a linear equation written in standard form. Tell students to suppose that they have 250 ft of fencing. Then demonstrate that the equation $2l + 2w = 250$ shows the dimensions of all the possible rectangular areas that could be fenced in with 250 ft of fencing.

Error Analysis
Students can use mental math to help them check their equations in exercises 7 and 8. When an equation is written in standard form, they can cover the x-term to find the value of y, which is also the y-intercept. (Make sure students understand that this is the same thing as substituting 0 for x.) For example, covering the x in exercise 7 gives $-2y = -4$, or $y = 2$. In exercise 8, covering $3x$ gives $3y = -9$, or $y = -3$. Therefore, students should know that the y-intercepts for the equations in exercises 7 and 8 are 2 and −3, respectively.

Some students may have difficulty getting started when changing an equation to slope-intercept form. Remind them that they need to get y by itself to the left of the equal sign.

Lesson 6•4 Graphing Absolute-Value Equations

Chalkboard Activity
Introduce students to absolute-value equations by letting them graph the values for x and y given in the example $y = |x - 3|$. Ask students to describe how this graph differs from others they have worked with in this chapter and the previous chapter. [It changes directions in straight lines.]

Using Manipulatives
The graph of an absolute-value equation always consists of a ray and its mirror image. The *vertex*, or the point at which the rays meet, will always be on the x-axis, but it will not always be point (0, 0). Ask pairs of students to plot the following points: (−1, 0), (0, 1), (1, 2), and (2, 3). Then suggest that each pair place a small mirror just to the left of the point (−1, 0). The ray and its mirror image represent the graph of $y = |x + 1|$.

Error Analysis
Students will choose their own x-values in exercises 6 and 7. After making a table of values and then graphing the points in the table, it is possible that students will not end up with a V-shaped graph. If this happens, suggest that they select several additional values for x. Students should always find the value of y when $x = 0$.

Final Check
Ask students if the graph of an absolute-value equation would ever dip below the x-axis. Be sure they can explain how they know. [The graph of an absolute-value equation cannot dip below the x-axis. The y-values are always positive or zero.]

Lesson 6•5 Other Graphs and Their Equations

Chalkboard Activity
Draw a downward-opening parabola. Then ask students to brainstorm a list of actions that could be represented by the curve. [Responses might include the path of a thrown ball or the path of a model rocket.]

Using a Calculator
Have students use a graphing calculator or a graphing utility on a computer to graph the equations in exercises 1 to 4. Have them discuss the slant of the lines and how it relates to the given equation. Be sure they understand that the slopes are the same in the equations for parallel lines and that the slopes are opposite reciprocals in the equations for perpendicular lines.

Working with ESL/LEP Students
Use your hands, rulers, or some other objects to illustrate the terms *parallel* and *perpendicular*. Use the same objects to illustrate lines that are not parallel and not perpendicular.

Name _____   Date _____

Chapter 6   Test

1. Find the slope of the line that passes through the points $(4, -1)$ and $(-6, 5)$. _____

2. Find the equation of a line with a slope of 3 and a y-intercept of -7. _____

3. Find the slope and y-intercept of the graph of the equation $12x - 3y = 36$. _____

Use the equation $8x = 12(y - 3)$.

4. Write the equation in slope-intercept form. _____

5. Write the equation in standard form. _____

Find the slope of each line.

6. A line parallel to $4x + 3y = 12$ _____

7. A line perpendicular to $3x + 2y = 6$ _____

8. Suppose between 1987 and 1994 the average annual individual contribution to Public Television went from $34 to $50. Find the annual rate of change in individual contributions. _____

Complete a table of values and then graph each equation.

9. $y = |x - 3|$

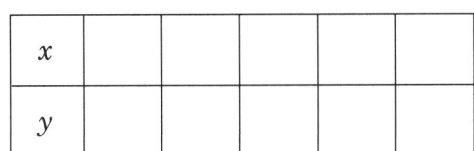

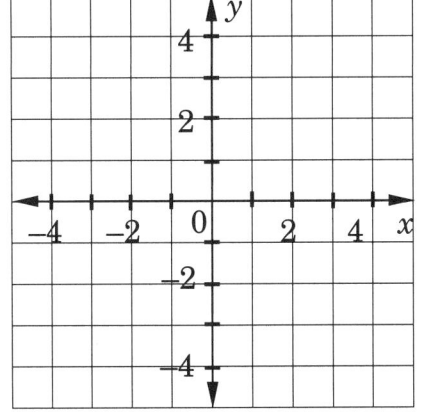

10. $y = x^2 - 1$

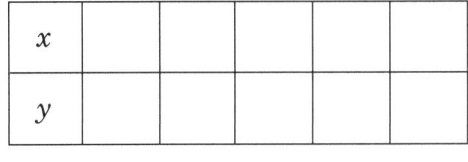

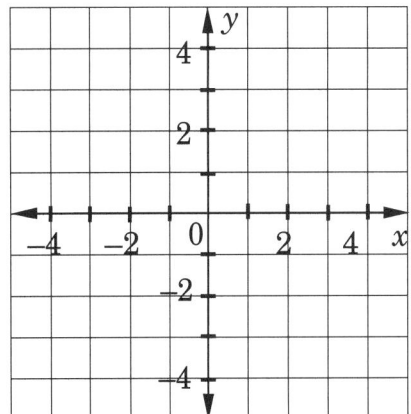

Chapter 6 Test   21

Chapter 7 Exponents

Introduction

Students might be interested in duplicating Galileo's experiments with falling objects. They could use a tennis ball and a basketball. Ask students to predict which ball will hit the ground first if both are dropped from the same height at the same time. Then have them test their predictions. [Both objects hit the ground at the same time.] Encourage students to investigate the laws of motion that govern the behavior of falling objects.

Lesson 7•1 More About Exponents

Chalkboard Activity
Continue with the example of doubling the thickness of paper. Students will see how the numbers increase quickly. You might also use a monetary example. If students started with $1 and doubled the amount every day, how much would they have in 30 days? [over $500,000,000]

Using a Calculator
Students will need to practice the proper key sequence for evaluating an exponential expression. Point out the x^y or y^x key on a scientific calculator and demonstrate how to use it to evaluate an expression such as 6^3 (some calculators have the x^y key as a 2nd function key): 6 x^y 3 $=$. Stress that the base is always entered before the x^y key and that the exponent is always entered after the x^y key.

Error Analysis
In exercises 11 to 14, some students may make the mistake of multiplying the base and the exponent. Ask them to evaluate pairs of expressions such as 3^{10} and 10^3. Students will recognize that something is wrong when they get the same value for both expressions.

Lesson 7•2 Writing Very Large and Very Small Numbers

Chalkboard Activity
Have students search through newspapers and magazines to find examples of very large and very small numbers. Then ask volunteers to write their numbers on the board and to indicate what their numbers were used to describe.

Error Analysis
Some students may think that the exponents in exercises 17 to 20 tell how many zeroes to add to the right of the decimal point. Stress that each exponent tells how many places to the right the decimal point should move.

Final Check
To assess students' understanding of scientific notation, discuss the following questions.

1. What three elements do all numbers written in scientific notation have in common? [They all have a number greater than or equal to 1 and less than 10, the multiplication symbol, and a power of 10.]

2. How is writing a very large number in scientific notation similar to writing a very small number in scientific notation? How is it different? [Students may say that both very large and very small numbers written in scientific notation have the three elements mentioned above. They may also say that very large numbers have positive exponents, whereas very small numbers have negative exponents.]

Lesson 7•3 Multiplication Properties of Exponents

Chalkboard Activity
Challeng students to write numbers written in scientific notation in standard form, and those in standard form in scientific notation.

1. 43,000,000 [$4.3 \cdot 10^7$]
2. $8.25 \cdot 10^7$ [82,500,000]
3. 0.00541 [$5.41 \cdot 10^{-3}$]
4. $7.9 \cdot 10^{-5}$ [0.000079]

Using a Calculator
Ask students to find the largest number that can be displayed on their calculators in standard form. Then have them explore ways of using scientific notation and the multiplication properties of exponents for computation with numbers greater than 99,999,999 or 9,999,999,999.

Alternate Teaching Approach

You might begin this lesson by pointing out that students have already had some experience using the product property of exponents. Draw a square on the chalkboard and label two sides x. A volunteer should be able to explain that the area of the square is $x \cdot x$, or x^2. Then you can demonstrate why $x \cdot x = x^2$.

$$x \cdot x = x^1 \cdot x^1 = x^{1+1} = x^2$$

Error Analysis

Some students may be confused about when to multiply exponents and when to add them. You may want to suggest that students change numbers written in exponential form to the products of factors. For example, they could rewrite exercise 3 as $(9 \cdot 9 \cdot 9 \cdot 9) \cdot (9 \cdot 9 \cdot 9)$ and exercise 4 as $(9 \cdot 9 \cdot 9 \cdot 9) \cdot (9 \cdot 9 \cdot 9 \cdot 9) \cdot (9 \cdot 9 \cdot 9 \cdot 9)$.

Lesson 7•4 Division Properties of Exponents

Chalkboard Activity

To review the multiplication of exponents, ask students to simplify each expression.

1. $8^2 \cdot 8^5$ [8^7]
2. $(8^2)^5$ [8^{10}]
3. $(2a)^2$ [$4a^2$]
4. $b^3 \cdot b^3$ [b^6]
5. $(x^2)^6$ [x^{12}]

Using a Calculator

Challenge students to simplify $\frac{7^5}{7^2}$ with the aid of a calculator. Suggest that they try using the grouping (parentheses) keys. Students' key sequences should be similar to the one shown below.

Working with ESL/LEP Students

Review the meaning of terms such as *base*, *exponent*, and *quotient*. Be sure students understand that the quotient property of exponents only applies to numbers with the same bases.

Error Analysis

Watch for students who try to divide the bases. In exercise 3, for example, they might simplify $\frac{5^5}{5^3}$ as 1^2. Point out that the base does not change. Also, make sure students understand that another way to write the expression in exercise 6 is $\frac{4^{11}}{4^1}$.

Lesson 7•5 Negative Exponents

Chalkboard Activity

To review division properties of exponents, challenge students to simplify each expression.

1. $\frac{6^7}{6^3}$ [6^4, or 1,296]
2. $\frac{5^3}{5}$ [5^2, or 25]
3. $\frac{9^4}{9^3}$ [9]
4. $(\frac{x}{3})^2$ [$\frac{x^2}{9}$]
5. $\frac{32r^5s^4}{8r^2s^2}$ [$4r^3s^2$]

Alternate Teaching Approach

You may want to begin the lesson by asking students to examine the following pattern.

$10^4 = 10,000$

$10^3 = 1,000$

$10^2 = 100$

$10^1 = 10$

$10^0 = 1$

$10^{-1} = \frac{1}{10}$

$10^{-2} = \frac{1}{100}$

$10^{-3} = \frac{1}{1,000}$

$10^{-4} = \frac{1}{10,000}$

Be sure students understand that the number on the right side of each equation is $\frac{1}{10}$ the number above it. Using the pattern, students should be able to predict that $10^{-5} = \frac{1}{100,000}$. Some students may wonder why $10^0 = 1$. A lengthy explanation is not necessary here because the meaning of a zero exponent is the topic of the next lesson.

Error Analysis

For exercises 10 to 13, students may need to be reminded that two separate steps are involved. In exercise 14, you may need to remind students that $7y^{-3} = 7 \cdot y^{-3}$.

Lesson 7•6 The Exponent of Zero

Chalkboard Activity

Write the following headings: "Product Property of Exponents," "Power Property of Exponents," "Power of a Product Property," "Quotient Property of Exponents," and "Power of a Quotient Property." Invite volunteers to write examples under the appropriate headings. You may want to write the equation $a^x \cdot a^y = a^{x+y}$ to provide students with the first example.

Chapter 7 Exponents 23

Using a Calculator

Have students use calculators to verify that any number they choose, no matter how large or how small, will yield 1 when raised to the zero power. You may need to remind them of the proper key sequence: the base is entered before the [x^y] key and the exponent is entered after the [x^y] key.

Alternate Teaching Approach

Finding patterns almost always makes mathematical theory easier to understand. Ask students to continue each of the following patterns.

$2^4 = 16$ $\quad$ $5^4 = 625$

$2^3 = 8$ $\quad$ $5^3 = 125$

$2^2 = 4$ $\quad$ $5^2 = 25$

$2^1 = 2$ $\quad$ $5^1 = 5$

$\underline{} = \underline{}$ $\quad$ $\underline{} = \underline{}$

Be sure they can explain that the exponents in each pattern decrease by 1, while each number on the right-hand side of an equation is $\frac{1}{2}$ or $\frac{1}{5}$ the number above it.

Error Analysis

Exercises 1 to 8 all have the same answer (1). Therefore, students may conclude, incorrectly, that the value of any expression that contains 0 as an exponent is 1. It may help to point out that $3k^0$ in exercise 9 can also be written $3 \cdot k^0$.

Lesson 7•7 Radicals

Chalkboard Activity

To prepare students for the idea that squaring and finding the square root are inverse operations, you might ask them to list pairs of inverse actions, or actions that undo each other. Examples include coming into and leaving a room, dropping a book and picking it up, and walking to school and walking home.

Using a Calculator

Students can use calculators to find square roots and cube roots, but they may need some help with the proper key sequence, especially for cube roots. On many calculators students will have to press [INV] or [2nd] first.

Alternate Teaching Approach

Ask students to refer to the illustrations of the square and the cube in Lesson 7•1. Point out that the dimensions given for the square are square roots and that the dimensions for the cube are cube roots.

Name _____ Date _____

Chapter 7   Test

Use a base and an exponent to write each product.

1. $4 \cdot 4 \cdot 4 \cdot 4 \cdot 4$ _____
2. $y \cdot y \cdot y \cdot y \cdot y \cdot y$ _____

Evaluate each expression.

3. 4^3 _____
4. $3^4 \cdot 3^2$ _____
5. x^0 _____
6. 5^{-3} _____

Write each number, using scientific notation.

7. 583,000,000,000 _____
8. 0.000135 _____

Write each number, using standard notation.

9. 5.1×10^4 _____
10. 7.63×10^{-3} _____

Simplify. Express your answers with positive exponents.

11. $3^6 \cdot 3^4$
12. $(x^3)^2$
13. $(2b)^3$

14. $6^{-4}(6^{-2})$
15. $\frac{x^{10}}{x^4}$
16. $\left(\frac{x}{3}\right)^4$

17. $\frac{60x^5}{20x^2}$
18. $\sqrt{y^6}$
19. $\sqrt[3]{27}$

Use the Pythagorean Theorem, $a^2 + b^2 = c^2$ to solve the following exercise.

20. A 25-ft ladder is leaning against a wall. If the ladder is 5 ft from the base of the wall, how far up the wall will the ladder reach to the nearest tenth of a foot?

_____

Chapter 8 Inequalities

Introduction
Discuss types of situations where we use terms such as less than or at least. Most often these are statements involving a range of numbers. Use the football example on page 131 as a springboard to a discussion of other objects or situations in which a range of values is acceptable, rather than an exact value. For example, the weight of a baseball bat must fall within a certain range. Students may have curfews that fall within a range of times. Sometimes a price range is set on gifts. Encourage as many examples as possible. Use the inequality symbols listed on page 131 to describe some of these examples as inequalities.

Lesson 8•1 Equations and Inequalities

Chalkboard Activity
Draw a number line on the chalkboard. Label it with integers −10 through 10. Explain that when comparing two numbers, the number to the right on the number line is the greater number and the number farther to the left is the lesser number. Invite students to graph a number greater than 2 and a number less than or equal to −2. Use one color of chalk for numbers greater than 2 and another color for those less than or equal to −2. Explain that on a number line, an open dot is used to show numbers greater than 2, but not including 2. A closed dot is used to show that a number is included. The lines to the right of 2 and to the left of −2 indicate that any numbers along those lines can be included. Challenge students to write an equality for the points they graphed.

Working with ESL/LEP Students
Provide several examples that use inequality symbols. Write the inequalities on the board as you say them. Encourage students to write or say them as you write. Use nonmathematical examples, such as the height of a certain tree is less than the height of another tree. In this case, draw the trees at appropriate heights on either side of the inequality.

Show students a rectangular box to be used to hold books of different heights. Invite students to measure each book and the height of the box to determine if each book would fit onto the box shelf. Draw a number line on the board and encourage students to mark the heights of the books and determine which ones would fit into the box. Mark the height of the box with a filled 0. Encourage students to write an equality, using the number line to describe the range of heights that fit onto the box shelf. This process can be repeated for other examples, such as the height of furniture that must fit through a door, or the size of carry-on baggage allowed on a plane.

Error Analysis
For exercises 6, 8, 9, and 10, students may mistakenly use a closed dot instead of an open dot. They may also forget to shade the arrow at the end of the number line. Point out that only the solutions to an inequality should be shaded on the graph of the inequality. Stress that all of the solutions should be shaded. The arrow at the end of a number line indicates that the line extends indefinitely. Since it represents all positive or negative numbers, the arrow needs to be shaded if those numbers are part of the solution set.

Final Check
To assess students' understanding of equations and inequalities, ask the following questions.

1. What is the difference between a closed dot and an open dot on the graph of an inequality? [A closed dot over a number means the number is part of the solution set; an open dot means the number is not part of the solution set.]

2. What would the graph of $3x > 12$ look like? [The graph would be a ray with an endpoint at 4; the endpoint would be a closed dot.]

3. Invite students to draw the inequality $-3 \leq n < 4$. [A closed dot at −3 and an open dot at 4. A segment is drawn between the two dots.]

4. Challenge students to write an inequality representing the statement, a ladder L is at least 3m, but must be shorter than 7m if it is to stand in a garage. [$3m \leq L < 7m$]

Lesson 8•2 Solving One-Step Inequalities

Chalkboard Activity
As an introduction to this lesson, write the following equation on the board.

$$4.56 - 0.01 \leq w \leq 4.56 + 0.01$$

Challenge students to give three equivalencies for w, to the nearest hundredth. [4.55, 4.56, 4.57] Discuss how this rule can be used to solve equalities and inequalities.

26 Chapter 8 *Inequalities*

Working with ESL/LEP Students

Students may have difficulty understanding why they must reverse the inequality symbol when multiplying or dividing both sides of an inequality by a negative number. Consider an inequality such as $3 > 2$ and suggest that students multiply each side by 4 and -4. Students will see that although it is true that $12 > 8$, it is not true that $-12 > -8$. Reversing the symbol makes the latter statement true. Invite students to draw a graph on a number line to represent each inequality.

Alternate Teaching Approach

Write the following problem on the board or read it aloud.

Ticket prices for a concert are $4, $5, $6, $7, and $8. Evon wants to buy two tickets for himself and a friend. He has $20, and $7 is needed for train fare. What combinations of tickets can he choose?

Discuss how this problem can be solved by adding combinations of tickets. Help students understand that it can also be solved as an inequality. Invite students to write an inequality statement for this problem and find the solution set. [$2c + 7 \leq 20$; c can equal two $4 tickets, or two $5 tickets, or two $6 tickets] Emphasize that the steps in solving the inequality are the same steps used in solving an equation.

Error Analysis

Students who have difficulty with exercise 13 may not be translating the inequality into an English expression correctly. Provide a translation, *20 times x is less than 100*. Ask students how they would write a word problem to represent the equation $20x = 100$. This may lead them to a word problem that represents the inequality.

Final Check

Review the following rules for solving inequalities.

1. The same number may be added or subtracted from each side of an inequality.

2. Each side of an inequality may be multiplied or divided by the same positive number.

3. Each side of an inequality may be multiplied or divided by the same negative number; remember to reverse the inequality.

Lesson 8•3 Solving Two-Step Inequalities

Chalkboard Activity

Review the rules for solving equations. Challenge students to solve the following equations for x.

1. $-3x + 2 = 14$
   [$x = -4$]
2. $\frac{x}{4} - 7 = -5$
   [$x = 8$]
3. $3x + 550 = 1{,}600$
   [$x = 350$]

Discuss how the same rules apply to solving inequalities. Provide students with examples of two-step inequalities, such as those listed below, and challenge them to find the solution sets. Remind them to reverse the direction of the inequality symbol when multiplying or dividing both sides of the inequality by a negative number.

1. $3x > 5x - 14$
   [$x < 7$]
2. $-5x > -115$
   [$x < 23$]
3. $-4x + 1 > 0$
   [$x < \frac{1}{4}$]

Working with ESL/LEP Students

Review the rules for solving inequalities while solving an inequality on the board. Invite students to graph the solution set on a number line. Challenge them to choose a number from the solution set and substitute it for the variable to verify that it is a true statement. Encourage them to substitute a number outside the solution set on the number line for the variable to see how it does not make a true statement. Provide students with examples of inequalities where the direction of the inequality symbol must be changed when multiplying or dividing by a negative number. Graph the inequality before and after changing the symbol to show why it needs to be changed to make a true statement.

Alternative Teaching Approach

Review the procedure for solving a two-step inequality. Before exploring the problem presented on page 136, challenge students to solve this problem. The total price for four concert tickets is $62.50, including a service charge of $2.50. What is the cost of one ticket? If necessary, help students set up and solve the equation. [$4x + 2.50 = 62.50$, $x = 15$ The price of one ticket is $15.] Explain why the procedure for solving two-step inequalities is very similar.

Error Analysis

Remind students that exercises 4 to 12 involve more than one step. Some students may forget to change the direction of the inequality symbol when multiplying or dividing by a negative number. Students may need to review the rules for multiplying and dividing by negative numbers.

Name _____   Date _____

Chapter 8 Test

Write the inequality that is graphed on the number line.

1.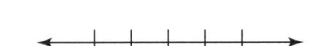

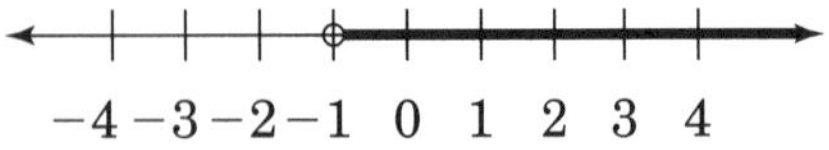

2.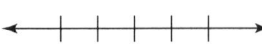

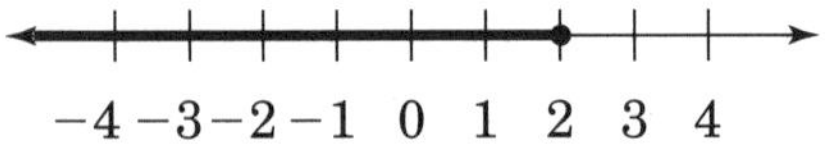

_____   _____

Label the number line. Then graph the inequality.

3. $x \geq 1$ 4. $x < 0$

State whether the inequality symbol *stays the same* or *should be reversed* in the solution of the given inequality.

5. $\frac{y}{4} < -2$ 6. $5x + 3 > 18$ 7. $-2x > 16$

_____   _____   _____

8. $\frac{x}{-6} > 4$ 9. $2x \geq 10$

_____   _____

Solve each inequality.

10. $x + 5 \leq 12$ 11. $-3k \geq 15$ 12. $18 < t - 7$

_____   _____   _____

13. $\frac{m}{-2} \leq 9$ 14. $8.7 > -3x$ 15. $8n + 3 > 4n - 1$

_____   _____   _____

16. $-x + 5 < 3x + 1$ 17. $5(x + 4) \geq 2(x - 5)$ 18. $x + 3x - 5x < -18$

_____   _____   _____

Solve.

19. Mary Jane makes more than $7.50 an hour. Write this expression as an inequality. _____

20. If Jamal deposits $1,500 in a bank, what interest rate must he get to earn at least $50 in interest? (Assume that interest is compounded annually.)

_____

28 Chapter 8 Test

Chapter 9 Monomials and Polynomials

Introduction
Start with the formula "A = lw" on the chalkboard and ask students if they remember how it is used. Brainstorm other formulas they recall, describing their uses. Ask students why variables and formulas are an important part of mathematics.

Lesson 9•1 Recognizing Monomials and Polynomials

Chalkboard Activity
Evaluate each of the following if $a = -1$, $b = 3$, and $c = 0.6$.

1. $(ab)c$ $[-1.8]$
2. $ab + ac$ $[-3.6]$
3. $b^a - c$ $[-0.266]$
4. $3bc - 2ab$ $[11.4]$
5. $5a - 3b + 4c$ $[-11.6]$

Ask students to write examples of the following:

6. An expression with two terms and one variable.
7. An expression with three terms and two variables.

Working with ESL/LEP Students
Students have been introduced to several new words. It will be helpful for students to review in their native language words such as *variable, exponent, terms,* and *expression*. Working in small groups, students could use formulas or equations and label examples of mathematical words. They could also make index cards with similar examples.

Alternate Teaching Approach
Ask students to name as many words as they can that contain the prefixes *mono-, bi-, tri-* and *poly-,* and give the meaning of these words. Some examples are: *monogram, biped, biceps, bicycle, tricycle, tripod,* and *polygon*. Help students connect these words to the terms *monomial, binomial, trinomial,* and *polynomial*. They may find it useful to make a poster of examples of these new terms with labels.

Error Analysis
Be sure students understand that xyz is a monomial with three variables, not a trinomial. Terms must be separated by operators (+ or −) for a polynomial to be a trinomial, so $x + y - z$ would be a trinomial.

Lesson 9•2 Adding and Subtracting Polynomials

Chalkboard Activity
Tell whether each expression is a monomial, binomial, trinomial, or polynomial, and state the degree. You may want to review the rules for determining degree.

	Type	Degree
1. $4m^2 + 5m - 8$	trinomial	2
2. $12x^2y$	monomial	3
3. $-6x^2 + 7x - 5x^3 + 12x^7$	polynomial	7
4. $8xyz + 4x^3 - 6x^4$	trinomial	4
5. $2x^3y + 5x^2$	binomial	4

Using Manipulatives
Invite students to use centimeter grid paper and cut 10 squares measuring 5 cm by 5 cm. Provide 10 centimeter cubes to each group. Tell students to call each paper square x by $x = x^2$.

Each centimeter cube can be named x^3. Have them add $2x^2 + 3x^3 + 4x^2 + 5x^3$ and find the answer. $[8x^3 + 6x^2]$ It should be obvious that the answer is not 14 paper squares or 14 centimeter cubes, but is 8 centimeter cubes and 6 paper squares, or $8x^3 + 6x^2$. Do a few more examples with these manipulatives, such as:

$4x^2 + 3x^2 + 5x^3 + x^3$ $[6x^3 + 7x^2]$ and
$x^2 + x^2 + 2x^3 + 8x^3$ $[10x^3 + 2x^2]$

Be sure students write the problems and record their answers graphically and algebraically.

Alternate Teaching Approach
Ask students to give examples of items that are sorted into like groups. Some examples might be the arrangement of tapes at a video store or the grouping of sodas in a soft drink machine. Discuss reasons for grouping like items. Help students connect their examples to combining like terms in an expression.

Lesson 9•3 Multiplying a Polynomial By a Monomial

Chalkboard Activity
Write the following expressions on the board. Invite students to combine like terms. Have them label the answer as a binomial, trinomial, or polynomial, and state the degree of the polynomial. You might provide a few items in the chart and let students fill in the rest.

	Answer	Type	Degree
1. $5y - 3y^2 + 4y + 2$	$[-3y^2 + 9y + 2]$	[trinomial]	[2]
2. $8x - 5 + 6x$	$[14x - 5]$	[binomial]	[1]
3. $3x^2y - 6x^2y + 12x^2y$	$[9x^2y]$	[monomial]	[2]
4. $2m + 3n - 4m + 8n$	$[-2m + 11n]$	[binomial]	[1]
5. $-3x^3 + 4x^2 - 2x$	$[-3x^3 + 4x^2 - 2x]$	[trinomial]	[3]

Alternate Teaching Approach
Multiplication can be done by modeling the area of a rectangle. This exercise will provide review of the meaning of area, combining like terms, multiplication, and the Distributive Property by giving concrete visual examples.

For example, $3a(4a + 5b)$ could be drawn as follows.

	4a	5b	
3a	I 12a²	II 15ab	3a
	4a	5b	

The area of rectangle I is $3a(4a) = 12a^2$ and the area of rectangle II is $3a(5b) = 15ab$, so the area of the combined rectangles is $12a^2 + 15ab$. The area of the big rectangle can also be found by multiplying the total length, $4a + 5b$, by the width, $3a$.

$3a(4a + 5b) = 12a^2 + 15ab$

Thus, you have illustrated that multiplication is distributive over addition.

Error Analysis
Be sure students understand that they should add exponents when multiplying variables with the same base. Remind them that $3x(2xy) = 6x^2y$, not $6xy^2$. Also remind them that $(2x)(2x) = (2x)^2 = 4x^2$, not $2x^2$.

Some students may try to apply the Distributive Property to an expression such as $3(4 \cdot 5)$ and get $3 \cdot 4 \cdot 3 \cdot 5$. Remind them that this property always involves two different operations, such as multiplication and addition (as in the geometric modeling above), or multiplication and subtraction.

Lesson 9•4 Multiplying Polynomials

Chalkboard Activity
To review the previous lesson, multiplying a polynomial by a monomial, ask students to find the errors in these products. Ask for volunteers to come to the board, indicate the error, and write the correct answer.

1. $x(x^2 - 4) = x^3 - 4$ $[x^3 - 4x]$
2. $7 - 3(5x + 9) = 7 - 15x + 9 = -15x + 16$
   $[7 - 15x - 27 = -15x - 20]$
3. $(8x^2 - 1)3x = 24x^3 - 3$ $[24x^3 - 3x]$
4. $-6(x - 3y) = -6x - 18y$ $[-6x + 18y]$

Alternate Teaching Approach
The idea of area can be extended to multiplying polynomials. You can connect geometry to algebra by using the areas of geometric figures to illustrate products of polynomials. Most students need this graphic approach to enhance their understanding of the method and result.

Example: $(x + 2)(3x - 5)$

	x	2	
3x	I $3x^2$	II $6x$	3x
-5	III $-5x$	IV -10	-5
	x	2	

The area of rectangle I is $3x^2$, II is $6x$, III is $-5x$, and IV is -10.

So, the area of the rectangle is
$(x + 2)(3x - 5) = 3x^2 + x - 10$.

Error Analysis
Sometimes, students confuse the rules for multiplying polynomials and adding like terms to give an incorrect sum, such as $4x^2 + 8x^2$ is equal to $12x^4$. This common error of adding exponents when adding coefficients can be avoided by using a picture to replace x^2. For example, use $x^2 = \triangle$. Then, using pictures, $4\triangle + 8\triangle = 12\triangle$, not 12 squares.

30 Chapter 9 *Monomials and Polynomials*

Name _____ Date _____

Chapter 9 Test

Tell whether each polynomial is a monomial, binomial, or trinomial. Then state the degree of the polynomial.

1. $5xy^2 - 3y$ _____
2. $4x^2 + 3x - 5$ _____
3. $8x^2y^5$ _____
4. $8x^4 - 5x^2 + 7$ _____

Simplify each expression.

5. $(2x - 5) + (3x^2 - 6x)$
6. $(x^2 + 4x - 3) + (2x^2 + 9)$

7. $(8a^2 + 3b) - (7b - 4a^2)$
8. $8(4x^2 - 5x + 2)$

9. $(3x^2 + 5x - 7) - (2x^2 - 4x)$
10. $6x(3x^2 - 4x + 2)$

11. $x^5(4x^2y^3)$
12. $(5x - 1)(4x + 3)$

13. $6xy(5x^2y^2 - 4xy + 6)$
14. $8x(4x^3 - 7x^2 + 9x - 3)$

15. $(5x + 2)(5x + 2)$
16. $(-3x + 1)(2x + 1)$

17. $5abc(3a - 6b)$
18. $(8x^2 + 3x - 5) - (4x^2 + 7x - 2)$

19. Find the volume of this rectangular prism. Simplify your answer. _____

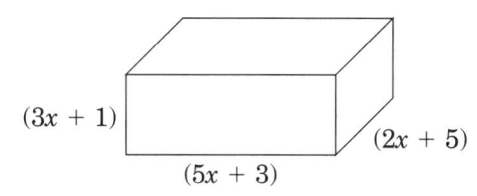

$(3x + 1)$, $(5x + 3)$, $(2x + 5)$

20. A room is x yards wide and $x + 2$ yards long. Write an expression that represents the cost to carpet the room if carpeting costs $25 per square yard. _____

Chapter 9 Test   31

Chapter 10 Factoring

Introduction
Provide examples of e-mail addresses in addition to that on page 157. Challenge students to break each address into parts. (If you need a source of e-mail addresses, look in newspapers and magazines; many publications encourage readers to send comments via e-mail.) Ask students whether they or anyone they know has an e-mail address. Invite those who do not already have an e-mail address to create one, using the addresses you provide as models.

Lesson 10•1 Finding the Greatest Common Factor

Chalkboard Activity
Write the following expressions and challenge students to find the products. Ask students to name the factors in each exercise.

1. $12 \cdot 3$ [36]
2. $(-4)(-8)$ [32]
3. $-5a(a2)$ [$-5a^3$]
4. $3x(5x)$ [$15x^2$]

Tell students that instead of finding products, they will find factors of monomials in this lesson.

Using Manipulatives
Have students work in small groups. Provide each group with a box containing an assortment of small objects such as coins, marbles, or attribute blocks. Then challenge them to devise a method for sorting the objects. Stress that, in order to sort the objects into groups, they must be able to identify qualities or characteristics that the objects have in common.

Alternate Teaching Approach
Reviewing the following divisibility rules may help students make their lists of factors.

- If the number is even, then 2 is a factor.
- If the number ends in 5 or 0, then 5 is a factor.
- If the number's digits add to a multiple of 3, then 3 is a factor.
- If 2 and 3 are factors, then 6 is a factor.
- If the last two digits are divisible by 4, then 4 is a factor.
- If 2 and 4 are factors, then 8 is a factor.
- If the digits of the number add to a multiple of 9, then 9 is a factor.
- If the last digit is zero, then 10 is a factor.

Error Analysis
Students may have trouble finding the greatest common factor when variables and exponents are present in the expression. Remind them that when factoring terms with variables and exponents, they should factor out the smallest exponent. For example, $5x^2y^3$ and $7x^3y^5$ have x^2y^3 as a common factor.

Lesson 10•2 Factoring Expressions

Chalkboard Activity
Ask students to find the greatest common monomial factor for each of the following.

1. $3a$, $3b$ [3]
2. $3x^2$, $5x$ [x]
3. $25x$, $5y$, $125z$ [5]
4. $12a^3b^2$, $8a^5$ [$4a^3$]

Alternate Teaching Approach
To help students factor trinomials, set up the two binomials for them, including the variables and signs. Students can then focus on finding the factors of the last term of the trinomial. Encourage students to try different combinations until they find the two factors that work. For example, for $x^2 - x - 20$, write $(x - \underline{})(x + \underline{})$, emphasizing that the signs must be a negative and a positive, since the middle and last terms are both negative. By factoring 20, students can see that 5 and 4 are the only factors that can be inserted into the two binomials to give $x^2 - x - 20$.

Error Analysis
All of the trinomials in exercises 15 to 20 can be factored. Ask students who find that one or more of these trinomials cannot be factored whether they are sure they have tried all the possible pairs of factors for their binomials.

Final Check
To assess students' ability to factor polynomials, discuss the following questions.

1. Is $8x^3$ the greatest common monomial factor of $16x^3 - 8x^2 + 32x$? Tell how you know. [No; the greatest common factor of the coefficients is 8 and of the variables is *x*. The greatest common monomial factor of the polynomial is 8*x*.]

2. When factoring polynomials, how can you check your answers? [Multiply the factors; if the product is the original polynomial, the answer checks.]

32 Chapter 10 *Factoring*

Lesson 10•3 Factoring Quadratic Expressions

Chalkboard Activity
As review, ask students to multiply the following binomials.

1. $(x + 3)(2x - 1)$ $[2x^2 + 5x - 3]$
2. $(3x - 5)(x - 2)$ $[3x^2 - 11x + 10]$
3. $(4x + 5)(2x + 3)$ $[8x^2 + 22x + 15]$
4. $(2x - 4)(3x + 5)$ $[6x^2 - 2x - 20]$

Alternate Teaching Approach
Point out that students already know how to factor $x^2 + bx + c$. For example, to factor $x^2 + 3x - 4$ students could first find all the factors of -4 and then determine which factors sum to 3.

Factors of -4	Sum of Factors
$2(-2)$	$2 + (-2) = 0$
$4(-1)$	$4 + (-1) = 3$ *
$-4(1)$	$-4 + 1 = -3$

Since $4 + (-1) = 3$, $x^2 + 3x - 4 = (x + 4)(x - 1)$.

To find the factors of $ax^2 + bx + c$, students can find all the factors of ac and see which of these sum to b. The first step in factoring $2x^2 + 5x - 12$, then, would be to find the factors of $(2)(-12)$, or -24.

Factors of -24	Sum of Factors
$1, -24$	$1 + (-24) = -23$
$-1, 24$	$-1 + 24 = 23$
$2, -12$	$2 + (-12) = -10$
$-2, 12$	$-2 + 12 = 10$
$3, -8$	$3 + (-8) = -5$
$-3, 8$	$-3 + 8 = 5$ *
$4, -6$	$4 + (-6) = -2$
$-4, 6$	$-4 + 6 = 2$

Since $(-3)(8) = -24$ and $-3 + 8 = 5$, $2x^2 + 5x - 12 = (2x - 3)(x + 4)$.

Final Check
To make sure students understand the concept of factoring polynomials, write $2x^2 + 2x - 4$ on the board. Ask them to identify the first step in factoring the polynomial. [Factor out a 2, to get $2(x^2 + x - 2)$.] Encourage students to tell how to finish factoring the polynomial. [Factor $x^2 + x - 2$ as the product of two binomials: $(x + 2)(x - 1)$. Therefore, $2x^2 + 2x - 4 = 2(x + 2)(x - 1)$.]

Lesson 10•4 Factoring Differences of Two Squares and Perfect Square Trinomials

Chalkboard Activity
As review, ask students to factor the following polynomials.

1. $a^2 + 5a - 14$ $[(a + 7)(a - 2)]$
2. $5s^2 + 23s - 10$ $[(5s - 2)(s + 5)]$
3. $2x^2 + 6x - 8$ $[2(x + 4)(x - 1)]$
4. $6v^2 - 16v + 8$ $[(2v - 4)(3v - 2)]$

Alternate Teaching Approach
Have students multiply the following binomials.

1. $(x + 3)(x - 3)$ $[x^2 - 9]$
2. $(x + 4)(x - 4)$ $[x^2 - 16]$
3. $(x + 5)(x - 5)$ $[x^2 - 25]$
4. $(2b + 3)(2b - 3)$ $[4b^2 - 9]$
5. $(2b + 4)(2b - 4)$ $[4b^2 - 16]$
6. $(2b + 5)(2b - 5)$ $[4b^2 - 25]$

Ask whether they notice a pattern. [When the factors are the same except for the signs, the resulting polynomial has no middle term.] Challenge students to find the product of $(x + 7)(x - 7)$ $[x^2 - 49]$ and $(2b + 6)(2b - 6)$ $[4b^2 - 36]$. Point out that the polynomials they have been exploring each have a difference of two squares. Explain to students that it is helpful to be able to recognize differences of two squares because they are always factored the same way. You may also want to use patterns to help students recognize and factor perfect square trinomials.

Error Analysis
Students may have difficulty factoring the polynomials in exercises 10 and 11. Some students may need to review the product property of exponents: $a^x \cdot a^y = a^{x+y}$.

Final Check
To assess students' understanding of factoring and the special polynomials covered in this lesson, discuss the following questions.

1. Is $3x^2(x^2 - 25)$ factored completely? If not, what do you need to do? [No, the binomial factor $x^2 - 25$ is a difference of two squares. Therefore, $3x^2(x^2 - 25) = 3x^2(x + 5)(x - 5)$.]

Chapter 10 *Factoring* 33

2. How can you factor $x^4 - 16$? [The expression is the difference of two squares: $(x^2 + 4)(x^2 - 4)$. One of the factors, $x^2 - 4$, is also the difference of two squares. Therefore, $x^4 - 16 = (x^2 + 4)(x + 2)(x - 2)$.]

Lesson 10•5 Zero Products

Chalkboard Activity
Ask students to solve the following equations.

1. $3x + 2 = 5$ [$x = 1$]
2. $4x + 1 = 0$ [$x = -\frac{1}{4}$]
3. $7x - 3 = 8x + 7$ [$x = -10$]

Tell students that they will be solving equations by using the zero product property in this lesson.

Using Manipulatives
The following game might help students understand the zero product property and remember that it can apply to more than two factors. Three-member teams work best. Ask each team member to select one of the factors in the equation $2x(x - 5)(x - 2) = 0$. Then have members take turns spinning a spinner whose sections have been labeled 0 to 5. (You may want to have students make their own spinners, using two pieces of cardboard and a thumbtack.) Each spin gives the value of x in $2x(x - 5)(x - 2) = 0$. Therefore, spins of 0, 2, or 5 will cause the equation to be true, that is, the solution is $x = 0$, $x = 2$, or $x = 5$. Members earn points whenever a spin causes the factor they selected to equal zero. After a predetermined number of rounds, members total their scores.

Final Check
Challenge pairs of students to create problems similar to exercise 11. One person provides the solution and the other writes the equation. The solution and equation can each be written on a separate piece of paper. Have partners exchange roles. Students can check their work by exchanging papers with other pairs.

Lesson 10•6 Solving Equations by Factoring

Chalkboard Activity
Review the previous lesson by having students write an equation for each solution. [Answers may vary. Possible answers are given.]

1. $x = 3$ or $x = -1$ [$x^2 - 2x - 3 = 0$]

2. $y = 0$ or $y = 4$ [$y^2 - 4y = 0$]
3. $b = 5$ [$b - 5 = 0$]

Using a Calculator
Students can use a calculator to solve quadratic equations. Enter the equation in the ($y=$) screen on a graphing calculator. Then graph the equation by pressing **graph**. Make sure that students have chosen an appropriate viewing window. They could use the trace key to locate the points where the graph crosses the x-axis. These are the points where $y = 0$. If their calculator has the capacity, they can use the Calc menu **2nd** **TRACE** 2 to find *roots*, which is another name for the zeroes of a function, or for the x-intercepts.

Error Analysis
Some students may have difficulty translating exercise 15 from words to an equation. You may want to remind them that *is* can be translated as $=$ and that *greater than* in this case can be translated as $+$.

Lesson 10•7 The Quadratic Formula

Chalkboard Activity
Review with students what they have learned by asking them to solve the following equations.

1. $x^2 + 2x + 1 = 0$ [$x = -1$]
2. $(c + 4)(c - 5) = 0$ [$c = -4$ or $c = 5$]
3. $4y(y - 3) = 0$ [$y = 0$ or $y = 3$]

Using a Calculator
Be sure that students understand what the $\pm$ symbol in the quadratic formula means. It does not mean that students should use the **+/−** key on their calculators; rather, the $\pm$ symbol tells students that they will need to find two answers.

Alternate Teaching Approach
To help students gain confidence with the quadratic formula, you might have them use it to solve some equations from the previous lesson. Because students have already solved these equations by factoring, they will easily be able to check their solutions when using the quadratic formula.

Error Analysis
Watch for students who do not realize that they cannot use the quadratic formula to solve the equations in exercises 3, 5, and 6, unless they first write the equations in the form $ax^2 + bx + c = 0$.

Name _____   Date _____

Chapter 10   Test

Write the greatest common factor of each group.

1. 12, 30, 48

2. $7x^2y$, $49x^2y^2$, $14xy^3$

Factor each polynomial.

3. $y^2 - 25$

4. $18r^2s - 45r^3s^3$

5. $x^2 + 8x + 16$

6. $a^2 - a - 30$

7. $x^2 - 5x + 6$

8. $2xy - xz$

9. $24a^2 + a^3 + 144a$

10. $64 + 25c^2 - 80c$

Solve each equation.

11. $y(y + 4) = 0$

12. $(a - 7)(a + 3) = 0$

Solve each equation by factoring.

13. $y^2 - 4y - 45 = 0$

14. $6a^2 - 7a = 10$

15. $y^2 + 2y = 63$

16. $12x^2 - 6 = x$

Use the quadratic formula to solve each equation.

17. $a^2 + 3a - 40 = 0$

18. $x^2 + 6x + 4 = 0$

19. $5x^2 + 3x = 2$

20. $2x^2 = 3x + 7$

Chapter 11 Systems of Equations

Introduction

List some specifics that might help detectives solve a mystery, such as a list of suspects, motives, evidence, witnesses, fingerprints, time of day mystery occurred, and so forth. Invite students to give examples how they have been a detective through their study of algebra. Ask them to list specifics they needed in order to solve various problems.

Lesson 11•1 What Is a System of Equations?

Chalkboard Activity

For a review of order of operations, write the following on the board. Encourage students to solve each equation by substituting the values of the variables listed below.

Evaluate each of the following if $a = -1$, $b = 3$, and $c = 2$.

1. $a^2 b$
   [3]
2. $5a - 3b$
   [−14]
3. $4c^2 - 5a^2$
   [11]
4. $(a - b)^3$
   [−64]
5. Are $5c - 2b$ and $2c - 5b$ equal?
   [No: $5(2) - 2(3) = 4$ and $2(2) - 5(3) = -11$]

Working with ESL/LEP Students

Clarify the term *system of equations* by explaining that a system is a group of things that are related or have something in common. Emphasize that a system of equations has the same variables. Introduce the use of equations to solve problems by using coins. Ask the following, "Mary has 7 coins made up of nickels and dimes. She has $.50 in change. How many of each coin does she have?" [4 nickels and 3 dimes] Encourage students to think of as many solution sets as possible.

Alternate Teaching Approach

Organize students into small groups and encourage them to make up problems dealing with their age and the age of a relative. When they have written the problem and solved it, ask them to trade problems with another group and see if they can solve the problems. Challenge the groups to find the problem most similar to the riddle that was carved on Diophantus' tomb.

Error Analysis

Students may have difficulty understanding the terms *less than, more than, sum of,* or *times that many.* Encourage students to divide story problems into small segments or units and write the mathematical expressions for each segment of the story before putting them together to formulate the story problem. Emphasize the importance of knowing what the story problem is asking and what information is necessary before it can be solved.

Final Check

Encourage students to find the solutions to exercises 7 and 8. Ask them to substitute the answer to verify the solution. This may also be a good time to review the order of operations: PEMDAS [parentheses, exponents, multiply, divide, add, subtract = **P**lease **E**xcuse **M**y **D**ear **A**unt **S**ally].

Lesson 11•2 Solving a System of Equations by Graphing

Chalkboard Activity

Draw a coordinate plane on the board. Then write the first equation. Invite students to solve for y if x equals 0, 1, 2, and 3. Challenge students to graph the solution set. Have students solve and graph the remaining equation on the board.

1. $3x + 4y = 12$

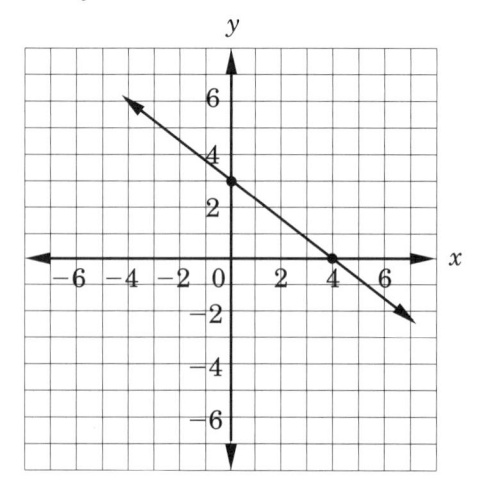

36   Chapter 11 *Systems of Equations*

2. $2x - 3y = 6$

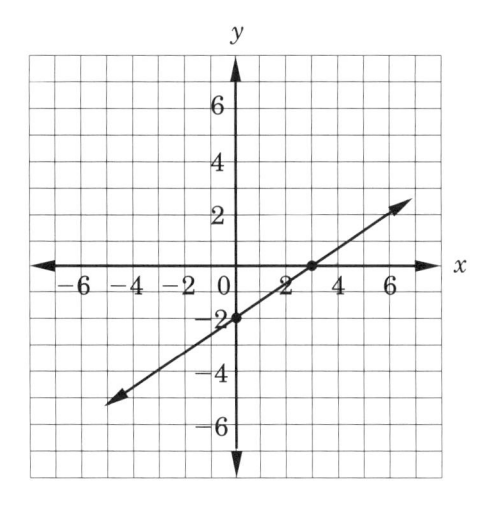

Using a Calculator
Encourage students to use a graphing calculator or a computer graphing program to graph both equations. The point of intersection of the two lines can be found by using either the trace key or the Intersect option if your calculator has one. For example, demonstrate the solution of this system of equations using a graphing calculator or a computer program: $y = 3x + 2$ and $y = -4x + 9$. [(1, 5)]

Error Analysis
Make sure students are using the order of operations correctly. While they are graphing their solution sets, emphasize the importance of plotting points using the correct axis.

Final Check
For exercise 9, invite students to write a system of equations for the story problem and graph the solution set. Encourage students to explain their answers to the class.

Lesson 11•3 Solving a System of Equations Using Addition or Subtraction

Chalkboard Activity
To review solving equations and substitution, write the following equations on the board. Invite students to solve the equations for the given variable. Then have them check their work by substituting the solution into the equation.

1. $6x + 7 = 25$ [$x = 3$]
2. $3x - 5 = 1$ [$x = 2$]
3. $4y = 24$ [$y = 6$]

Using a Calculator
Systems of equations can also be solved with the aid of a calculator. Students will need to write the equations in the form $y = ax + b$ in order to enter them into the calculator. Point out to students that even though the calculator can solve equations, it is not always the most efficient way. Some problems are actually easier to solve by using the methods of this lesson. You might use exercise 3 as an example. Adding the equations yields $24x = 48$, solving $x = 2$. Using a calculator, an exact answer usually takes longer because the zoom and trace functions take time to find the correct solutions.

Error Analysis
Students may get careless adding and subtracting positive and negative numbers. Emphasize the importance of paying attention to the signs preceding numbers. To minimize errors, take 5 minutes before the lesson for oral exercises, such as $5 - 10$.

Final Check
Remind students to check their solutions for each equation. Sometimes they arrive at a solution that only works in one of the equations.

Lesson 11•4 Solving a System of Equations Using Multiplication

Chalkboard Activity
For review of the Distributive Property, write the following on the board.

Multiply.

1. $2(a + 6b)$ [$2a + 12$]
2. $3(x - 3y)$ [$3x - 9y$]
3. $2(-2x - 4x)$ [$-12x$]
4. $4(-x + 5y)$ [$-4x + 20y$]
5. $3(5x + 6y)$ [$15x + 18y$]

Tell students they will be using the Distributive Property to solve a system of equations in this lesson.

Using Manipulatives
Have students work in small groups. One member of each group takes some coins from their pocket or purse. They should tell the rest of the group which coins and the total number of coins they have (limit them to just two types of coins, for example, dimes and quarters). They should ask the group to guess how many of each coin they have. Elicit from the class the fact that if they are only told how many

Chapter 11 *Systems of Equations*   37

coins and nothing else, they cannot solve the problem. Knowing how much money they have in addition to the total number of coins enables the rest of the group to find out how many of each coin they have. Let the next person in the group tell how many coins and how much money they have. Encourage the rest of the group to write equations describing the given information. Ask them to formulate a system containing two equations and explain how they found the solution. Repeat until everyone gets a chance.

Working with ESL/LEP Students

Pair students with English speaking students. Have one student put on the table any amount of money, limited to two types of coins. Instruct students to write two equations about the coins. The first equation tells the total number of coins; the second equation tells the value of those coins. Students write the two equations, using words to describe the situation, then introduce variables to describe the same thing. For example, suppose there are 3 dimes and 5 pennies on the table.

Number of coins: 3 dimes + 5 pennies = 8

Value of coins: 3 • 10¢ + 5 • 1¢ = 35¢.

Now students can work together to use variables to describe this same situation. Let d = the number of dimes (we know $d = 3$). Let p = the number of pennies (we know $p = 5$). Have them write two equations for the number of coins. [$d + p = 8$, $10d + 1p = 35$] Students can solve this system to check that they get the known number of coins.

Error Analysis

Students may multiply both equations by the same multiplier instead of only one equation to make the coefficients of a variable the same. This does not allow them to solve for one variable. When adding and subtracting equations, students may make errors in using this operation with positive and negative numbers. Encourage them to watch the signs preceding the numbers carefully. Emphasize that all solutions should be checked in both equations.

Lesson 11•5 Solving a System of Equations Using Substitution

Chalkboard Activity

To review substitution, write the following expressions on the board. Invite students to simplify the problems and explain how they arrived at their answers.

Simplify.

1. $3x + 12$, when $x = 3$ [21]
2. $5x - y$, when $x = 3, y = 4$ [11]
3. $3x - 3y$, when $y = x$ [0]
4. $x + 2y$, when $y = x + 4$ [$3x + 8$]
5. $-x - y$, when $y = x - 1$ [$-2x + 1$]

Tell students that in this lesson they will solve a system of equations, using substitution.

Alternate Teaching Approach

Explain that sometimes the form of the equation dictates the method of solution. Write these equations on the board: $y = 4x - 1$ and $y = 3x + 2$. Explain that graphing may be an efficient method if the answers are whole numbers or integers. The substitution method can also be used. Present to the class the premise that since both of the lines must pass through the same point to have an intersection, both of the y-values must be the same. Therefore, $4x - 1 = 3x + 2$ and in solving for x, $x = 3$. Substitute 3 for x in the first equation and encourage the class to find that $y = 11$.

Final Check

To assess students' understanding of solving systems of equations, ask the following questions.

1. When are you more likely to use the substitution method for solving a system of equations? [when a variable in one of the equations does not have a leading coefficient, or one of the equations is in the form $y = mx + b$]

2. Given the system $3x - 2y = 5$ and $x + 2y = 7$, which method are you more likely to use to solve the system? [addition: the coefficients of the y variable have opposite signs]

Name _____ Date _____

Chapter 11 Test

Tell whether (1, −3) is a solution of the system of equations.

1. $4x + y = 1$
   $2x - 3y = 11$ _____

Solve each system of equations by graphing.

2. $x + y = 8$
   $y = 3x$ _____

3. $3x + y = 5$
   $x = 2y + 4$ _____

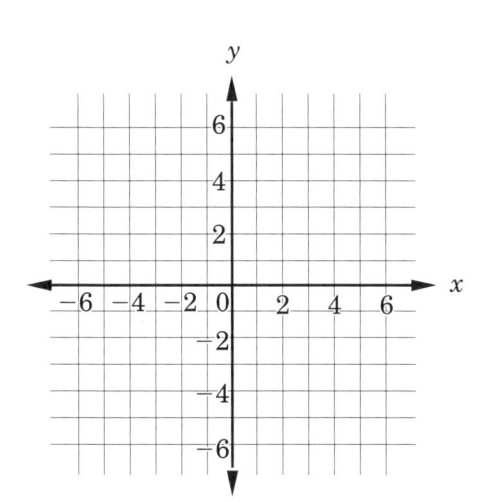

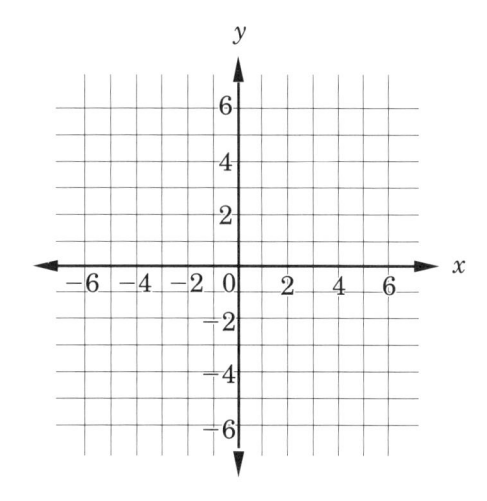

Solve the system of equations, using addition or subtraction.

4. $y = x + 3$
   $2x - y = 9$ _____

5. $x + 2y = 7$
   $x - 2y = -1$ _____

Solve each system of equations, using the multiplication method.

6. $3a + 3b = 15$
   $2a + 6b = 22$ _____

7. $x + 2y = 1$
   $-2x + 3y = 12$ _____

Solve each system of equations, using the substitution method.

8. $y = 2x - 1$
   $x + 3y = 25$ _____

9. $y - x = 4$
   $x + 2y = 2$ _____

10. One day a store sold 35 shirts. Blue shirts cost $15.95 and red shirts cost $20. How many of each color were sold if the store took in $619? _____

Chapter 11 Test 39

Answers

*Accept other reasonable answers when a possible answer or drawing is indicated.

Chapter 1 The Language of Algebra

1•1 Understanding Variables and Expressions

1. **a.** 1800
   b. Yes. Writing the number and variable next to each other is another way to represent multiplication, so 150 • n will have the same value as 150n.
2. 13 **3.** 8 **4.** 35 **5.** 15 **6.** 23 **7.** 1 **8.** 130 **9.** 3
10. 19
11. **a.** Yes. Multiplying 6 and h represents what the total pay will be. Subtracting 9 will show Donna's pay after her taxes have been deducted.
    b. $28.50
12. Possible answer: A variable is a letter used to represent a number. A variable expression is a mathematical sentence that includes variables, numbers, and operation signs.

1•2 Understanding Exponents

1. 2 • 2 • 2 • 2 **2.** 7^2 **3.** 6^4 **4.** 9^6 **5.** 12^3 **6.** 2^7
7. 8.5^2 **8.** 4 • 4 • 4 **9.** 10 • 10 **10.** 25 • 25 • 25
11. 3 • 3 • 3 • 3 • 3 • 3 **12.** 2 • 2 • 2 • 2 • 2 **13.** 97
14. 3 • 3 • 3 • 3 **15.** 6 • 6 • 6 **16.** 4.5 • 4.5 **17.** 81
18. 15 **19.** 2^{13}
20. **a.** 100 1,000 10,000 100,000 **b.** The number of zeroes in the answer is equal to the value of the exponent.

1•3 Using Grouping Symbols

1. (44 − 4) • 4 • 2 = 320, whereas 44 − (4 • 4 • 2) = 12
2. 5 + 3 **3.** 3 + 4 **4.** 600 + 10 **5.** $a + b$ **6.** $n + 12$
7. 16 − 9 **8. a.** 3 **b.** 39 **9. a.** 6 **b.** 26 **10.** 1
11. 77 **12.** $\frac{5}{8}$ **13.** 251 **14.** 16 **15.** 20
16. [5 + (5 ÷ 5)] • 5
17. Possible answer: 6 − (4 + 2)
18. 4(15) + 6(12); $132
19. Possible answer: Grouping symbols allow you to change the priority of calculations, and this can be done more than once in an expression.

1•4 Understanding Order of Operations

1. Division **2.** Multiplication **3.** Addition
4. Raise to power **5.** Division **6.** Raise to power
7. 10 **8.** 5 **9.** 24 **10.** 49 **11.** 19 **12.** 42 **13.** 5
14. 2 **15.** 15 **16.** 15 **17.** (2 + 5) • 2
18. 20 ÷ (15 ÷ 3) **19.** (12 + 20) ÷ 4 − 5
20. (14 − 5 − 2) • 2 **21.** 4 • 6 + 7 • 6 − 2 • 3; $60
22. Possible answer: Yes. If the calculator did not follow the order of operations, then the answer would have been 20.

1•5 Using Inverse Operations

1. Subtraction; addition **2.** Division; multiplication
3. Multiplication; division **4.** Addition; subtraction
5. Division **6.** Subtraction **7.** Multiplication
8. Addition **9.** Multiplication and addition
10. Division and subtraction
11. **a.** 30 **b.** Multiply 30 taxis by 4 tires each to see if this gives 120 tires.
12. **a.** Divide $75.60 by 6 people to figure out how much each person pays; $12.60 is the correct answer. **b.** Yes. To check their answer they multiplied $12.60 by 6 to get $75.60.

1•6 Verbal Expressions and Algebraic Expressions

1. • or ( ) **2.** ÷ **3.** − **4.** $k − 6$ **5.** $p ÷ 2$ **6.** $a + 3$
7. $\frac{d}{5}$ **8.** $2y$ **9.** $(x + y) ÷ 5$
10. Possible answer: Eight classmates sold popcorn at a fund-raiser. Each person spent b cents on supplies and sold the popcorn for a cents. What was their total profit?

Chapter 1 Review

1. variable **2.** evaluate **3.** exponent
4. parentheses; brackets; vinculum
5. 7 **6.** 3 **7.** 7^3 **8.** b^5 **9.** 12 • 12 • 12 • 12
10. $d • d$ **11.** 16 **12.** 1,000 **13.** 625 **14.** 9.6 **15.** 0
16. 6 **17.** 16 **18.** 3 **19.** 4 **20.** 18 **21.** Multiplication
22. Addition **23.** $n − 2$ **24.** $n − k$ **25.** $15 + \frac{1}{2}n$
26. $m + 8$ **27.** $2n − 7$
28. Possible answer: When riding a bicycle, stop signs and painted lines for crosswalks are symbols that make riding safer.

Chapter 1 Practice Test

1. 14 2. 15 3. 30 4. 9 5. 5 6. 13 7. h^3 8. 81
9. 144 10. 8 11. 21 12. 7 13. 7 14. 28 15. 32
16. Multiplication 17. Addition 18. Multiplication
19. $3c - 8$ 20. $d + 1.25$

Chapter 1 Test

1. 24 2. 48 3. 22 4. 36 5. 19 6. 11 7. m^5
8. 64 9. 256 10. 10 11. 84 12. 28 13. 13
14. 33 15. 65 16. Subtraction 17. Division
18. Addition 19. $2w + 5$ 20. $(3 + 3 + 3 + 8)d$

Chapter 2 The Rules of Algebra

2•1 Recognizing Like Terms and Unlike Terms

1. $6y$ 2. $3n$ 3. $15hp$ 4. wd 5. 2 6. 2 7. 3 8. 3
9. $5x, 6y$ 10. $k, 3d^4$ 11. Unlike 12. Unlike
13. Unlike 14. Unlike 15. Yes 16. Yes 17. No
18. Yes 19. 30, 18, 9
20. Like terms have the same variable parts. Unlike terms have different variable parts.

2•2 Understanding the Properties of Numbers

1. 9 2. 1 3. 9 4. 15
5. Commutative Property of Multiplication
6. Identity Property of Addition
7. Associative Property of Addition
8. Identity Property of Multiplication
9. Associative Property of Multiplication
10. Commutative Property of Addition;
    $\$4.25 + \$1.75 + \$1.29 = \$7.29$
11. Possible answer: Commutative Property of Multiplication means that the order of two factors does not change the product. In a 3×5 rectangle, the area can be found by 3×5 or 5×3.

2•3 Understanding the Distributive Property of Multiplication Over Addition

1. 6 and 9 2. 4, 3, 2 3. 8, x 4. 7, 7 5. 3 6. d, am
7. $3 \cdot 9 + 3 \cdot 4$ 8. $15 \cdot 8 + 15 \cdot 7$ 9. $8a + 8b$
10. $12.7d + 12.7w$ 11. $4(x + y)$ 12. $13(k + ab)$
13. 1,428
14. Use the Distributive Property to rewrite $6(a + b)$ as $6a + 6b$.

2•4 Adding Like Terms

1. Yes 2. No 3. Yes 4. Yes 5. No 6. Yes 7. $9p$
8. $26y$ 9. $9n$ 10. $33a + 6b$ 11. $16c + 11$
12. $7x + 7y$ 13. $14k + 3n + 37$ 14. $17a + 20.5b$
15. a. $12m + 9n$; 15. $15m + 7n$ b. $27m + 16n$
16. Add like terms. $5m + 7m = 12m$. Since $9m^2$ is not a like term, your final answer is $9m^2 + 12m$.

2•5 Subtracting Like Terms

1. No 2. Yes 3. Yes 4. No 5. Yes 6. Yes 7. $4g$
8. u 9. $15w$ 10. 0 11. $4.9a$ 12. $1\frac{1}{6}t$ 13. $4k$
14. $11a - 5a^2$ 15. $17h^2 + 25h + 9$
16. $0.6ab + 6.4a^2b + 1.7ab^2$ 17. $7x^2$ 18. $80r$
19. Answers may vary but should include the fact that $5a + 4a = 9a$.

2•6 Multiplying Terms

1. Yes 2. Yes 3. No 4. Yes 5. No 6. Yes 7. $28ab$
8. $15xy$ 9. $36mn$ 10. $56p^2$ 11. $20k^2$ 12. $28hk$
13. $3c^2$ 14. a^2b^2 15. $21st^2$ 16. $36mn$ 17. $30h$
18. $24x^2yz$ 19. $14c$ 20. $4c$ 21. $45c^2$
22. $4{,}320mx$ dollars
23. Multiply 5 times 7, times h, times w. The answer is $35hw$.

2•7 Dividing Terms

1. $\frac{24}{6}$ or $\frac{8}{2}$ 2. $\frac{18}{2}$ 3. $\frac{x}{y}$ 4. $\frac{3x}{5}$ 5. $\frac{12n^2}{2n}$ 6. $\frac{abc}{ab}$ 7. $6h$
8. 20 9. $12k$ 10. 3 11. c 12. 1 13. $2a$ 14. 3
15. $5m$ 16. b^2c 17. $6c$ 18. a. 32 b. $4n$
19. Possible answer: Multiply the answer by the divisor.

Chapter 2 Review

1. Distributive Property of Multiplication Over Addition
2. coefficient 3. factors
4. Terms: $2x, 3y, 5z$; coefficients: 2, 3, 5
5. Terms: $4m^3, 7m^2, 9.2m, 16$; coefficients: 4, 7, 9.2, 16
6. Unlike 7. Like 8. Like 9. Unlike 10. $20k$
11. $9x$
12. $2n$ 13. $9h$ 14. $35e$ 15. $7a + 5b$ 16. $8n + 3m$
17. $6p$ 18. $10y^2$ 19. $6p^2$ 20. $9ab + 4a + 13$
21. $10x^2y + 8x + y$ 22. $40ac$ 23. $3d^2v^2$ 24. $30m^2n$
25. $5b$ 26. $8xz$ 27. 1 28. $16n, 8n, 48n^2$, 3
29. Identity Property of Addition
30. Commutative Property of Multiplication
31. Associative Property of Addition
32. Commutative Property of Addition

Answers 41

33. $6x + 6y$ 34. $6a^2 + 8ab + 10ac$
35. a. $3d, 5d$ b. $8d$
36. Other activities might include time spent on homework or number of calories consumed.

Chapter 2 Practice Test

1. $3x^2, 2x, 5pr$ 2. $3, 2, 5$ 3. $8t$ 4. $8x$ 5. $5x$ 6. $17de$
7. $6n$ 8. 0 9. $12a + 2b + 11$ 10. $21x^2 + 26xy + 4y^2$
11. $21a$ 12. $12xy$ 13. $6k^2$ 14. $10x^2y$ 15. $6a$ 16. 32
17. x 18. $11bc$ 19. $18y, 12y, 45y^2, 5$
20. Associative Property of Multiplication
21. Identity Property of Multiplication
22. Commutative Property of Addition
23. $4a + 4b$ 24. $2x^2 + 6xy$
25. a. $6d^2$ yd^2 b. $\$150d^2$

Chapter 2 Test

1. $5y^3, 4x, 3nq$ 2. $5, 4, 3$ 3. $3m$ 4. $13r$ 5. $9y$
6. $2cg$
7. $15k$ 8. $2a$ 9. $12x - 5y - 15$
10. $4x^2 - 13xy + 11y^2$
11. $24b$ 12. $20xy$ 13. $21a^2$ 14. $9x^2y$ 15. $4y$
16. 17 17. y 18. $4ac$ 19. $30z; 18z; 144z^2; 4$
20. Commutative Property of Addition
21. Associative Property of Multiplication
22. Identity Property of Multiplication
23. $12a + 18b$ 24. $3x^2 + 6xy$
25. a. $10n^2$ in.^{2} b. $14n$ in.

Chapter 3 Equations and Formulas

3•1 What Is an Equation?

1. Yes; $4(5) = 20$ 2. False 3. Open 4. False
5. True 6. 11 7. 4 8. 18 9. 3 10. 4 11. 11
12. 7 13. 0 14. 36 15. 3.5 16. 2 17. 4
18. a. $1.5w = 9$ b. $\$6.00$
19. Possible answer: Substitute the value for the variable into the equation. If the equation is true, then the number is the solution to the open sentence.

3•2 Subtracting to Solve Equations

1. Subtract 5 from both sides.
2. Subtract 14 from both sides.
3. Subtract 8 from both sides.
4. Subtract 2 from both sides.
5. Subtract 13.7 from both sides.

6. 9 7. 3 8. 5 9. 3.5 10. 47 11. 432
12. a. $35 = 29.95 + d$ b. $\$5.05$
13. Possible answer: Have someone place 5 counters into the envelope on one pan of the balance. Place 9 counters on the other pan. Now remove 4 counters from both pans. The number of counters (5) remaining on one side of the balance should be the same number of counters in the envelope.

3•3 Adding to Solve Equations

1. Add 3 to both sides.
2. Add 7.5 to both sides.
3. Add 5 to both sides.
4. 8 5. 18 6. 20 7. 8.3 8. 23 9. 388
10. a. $l - 3.26 = 0.13$ b. 3.39 cm
11. Possible answer: After the first week of school, 3 children left the class. This left 10 children in the class. How many children were in the class at the start of the school year?

3•4 Dividing to Solve Equations

1. Dominick divided both sides by 8. Yes, he is correct.
2. Divide both sides by 4.
3. Divide both sides by 7.
4. Divide both sides by 2.5.
5. Divide both sides by 12.
6. Divide both sides by 4.
7. 4 8. 9 9. 6 10. 9 11. 12 12. 13 13. 1.5
14. 30 15. 3.6 16. $\frac{1}{21}$ 17. $\frac{4}{3}$ 18. $\frac{6}{5}$ 19. 2 20. 9
21. $\frac{1}{15}$ 22. $\frac{3}{8}$ 23. 36 24. 54 25. 18.4 26. 68
27. a. $\frac{2}{3}s = 7.6$ b. 11.4 miles per hour
28. Possible answer: Since multiplying by the multiplicative inverse of a number is the same as dividing by the original number, and $\frac{1}{2}$ is the multiplicative inverse of 2, dividing by 2 must be the same as multiplying by $\frac{1}{2}$.

3•5 Multiplying to Solve Equations

1. Multiply both sides by 3.
2. Multiply both sides by 7.
3. Multiply both sides by 3.1.
4. Multiply both sides by 5.
5. Multiply both sides by 2.
6. 12 7. 45 8. 30 9. 84 10. 33 11. 18 12. 24.92
13. 48 14. 561 15. 136 16. 0.621 17. 23.7
18. Possible answers: Use cross products, multiply both sides by 12, multiply numerator by 3
19. 21 20. 15 21. 40 22. 17.5 23. 6 24. 8
25. a. $4 = \frac{v}{25}$ b. 100 volts

26. Possible answer: Since 4 is added, subtract 4 from both sides. Since x is divided by 2, multiply both sides by 2.

3•6 What Is a Formula?

1. $20 = l(2)$ 2. $300 = 50t$ 3. $34 = (2)(12) + 2w$
4. $s = 84 - 16$ 5. $96 = 1.8C + 32$ 6. $36 = (\frac{1}{2})(8)h$
7. $26 = r - 5$; 31 8. $45 = 5r$; 9
9. $A = (2.5)(1.8)$; 4.5 10. $F = (1.8)(55) + 32$; 131
11. $12 = (\frac{1}{2})(8)h$; 3
12. a. 120 ft² b. 720 bricks c. $540
13. Possible answer: A formula is also an equation because it gives the relationship between two or more variables.

3•7 Solving Equations With Fractions and Mixed Numbers

1. Multiply by $\frac{6}{5}$, the multiplicative inverse of $\frac{5}{6}$.
2. Add $2\frac{15}{16}$. 3. Multiply by 7. 4. Subtract $\frac{2}{5}$.
5. Multiply by $\frac{3}{5}$. 6. $3\frac{1}{8}$ 7. $67\frac{1}{2}$ 8. 126 9. $3\frac{7}{12}$
10. a. $2\frac{1}{4} + 1\frac{7}{8} = x$; $4\frac{1}{8}$ pounds of flour
    b. $0.81 \div 2\frac{1}{4} = p$; $.36 per pound
11. $\frac{3}{7}x = 72$; 168 patients; Possible answer: Multiply both sides by $\frac{7}{3}$, the multiplicative inverse of $\frac{3}{7}$, to get 168 patients.

Chapter 3 Review

1. equation 2. algebraic sentence
3. multiplicative inverse 4. proportion; ratios
5. solution 6. Open 7. False 8. True 9. Open
10. 16 11. 24 12. 35 13. 8 14. 64 15. 6
16. 21.75 17. Subtraction 18. Division 19. $\frac{8}{3}$
20. $\frac{1}{24}$ 21. 60 22. 153 23. $4\frac{1}{2}$
24. Possible answers: $\frac{1.5}{5} = \frac{m}{2.5}$; $\frac{3}{4}$ cup of milk; $\frac{3}{5} = \frac{f}{2.5}$; $1\frac{1}{2}$ cups of flour
25. Other activities might include finding your rate walking to school or finding your average in math class.

Chapter 3 Practice Test

1. True 2. Open 3. 3 4. 2.5 5. 55 6. 2.5
7. Multiplication 8. Addition 9. $\frac{1}{8}$ 10. $\frac{9}{5}$ 11. 1
12. $\frac{4}{13}$ 13. 6 14. $1\frac{11}{12}$ 15. 72 16. 28.2 17. 210
18. 264 19. $p = 600 + 1{,}200$; $1,800
20. a. $d = rt$ b. $20 = \frac{4}{9}r$ c. 45 miles per hour

Chapter 3 Test

1. True 2. Open 3. 5 4. 2.5 5. 35
6. 7 7. Division 8. Subtraction 9. $\frac{8}{1}$ 10. $\frac{5}{7}$
11. $\frac{1}{10}$ 12. $\frac{4}{7}$ 13. 12 14. $2\frac{3}{4}$ 15. 68 16. 7.5
17. 120 18. 244.8 19. $12x = \$1{,}500$; $125 per month
20. a. $d = rt$ b. $37\frac{1}{2} = r(\frac{3}{4})$ c. 50 miles per hour

Chapter 4 Integers

4•1 What Is an Integer?

1. The quarterback lost 4 yards.
2. The temperature is 13 degrees above zero.
3. The price of CDs went down $1.50.
4. The balloon dropped 250 feet.
5. Len spent $200 during April.
6. -3 7. 6 8. 400 9. -15
10. 2 11. -20
12–14. Possible answers: Temperatures, or changes in temperature; account balances, or changes in account balances; yards gained or lost in football
15. 1,400; Possible answer: A positive integer because the point is above sea level
16. $-28{,}232$; Possible answer: A negative integer because the point is below sea level
17. Possible answer: Experienced recreational scuba divers can safely reach a depth of 130 feet below the surface (-130).

4•2 Integers and the Number Line

1. Yes. Positive 3 is three units to the right of zero and negative 2 is two units to the left of zero.
2. 2 3. -7 4. 0 5. 9 6. -4 7. -9 8. T 9. F
10. K 11. Z 12. J 13. P
14–17.

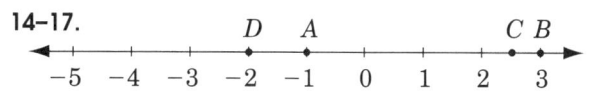

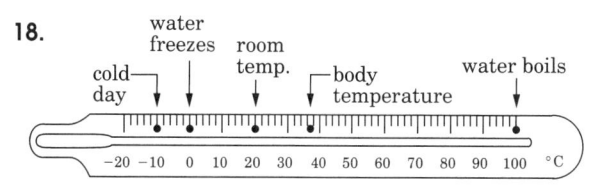

```
         D   A                    C B
◄──┼───┼───┼───┼───┼───┼───┼───┼───►
  -5  -4  -3  -2  -1   0   1   2   3
```

18. [thermometer diagram with labels: cold day, water freezes, room temp., body temperature, water boils; scale from -20 to 100 °C]

19. A change of 1°C is greater than a change of 1°F. Possible answer: Since there are 180°F between freezing and boiling, and only 100°C between freezing and boiling, each Fahrenheit degree represents a smaller unit of temperature.

Answers 43

4·3 Absolute Value

1. Yes. If a number is negative, its absolute value equals its opposite, which is a positive number.
2. 4 3. 3 4. 6 5. 0 6. 8 7. 10 8. 7 9. 5 10. 18
11. $4\frac{1}{2}$ 12. 19.6 13. 2.7 14. 7 15. 5 16. 18 17. 9
18. 35 19. 63 20. $-13°F$ 21. $22°F$
22. Possible answer: Distance from the ocean's surface is given by the absolute value of the altitude or depth. The absolute value of 1,220 is 1,220. The absolute value of $-1,450$ is 1,450. Since 1,450 is greater than 1,220, the submarine is further from the ocean's surface.

4·4 Which Integer Is Greater?

1. San Diego, Miami, Philadelphia, Atlanta. Found the percentage for each city on the number line, then listed the cities from right to left.
2. 6 3. 7 4. 0 5. -3 6. $<$ 7. $<$ 8. $<$ 9. $>$
10. $>$ 11. $<$ 12. $>$ 13. $=$ 14. $-4, -2, 3$
15. $-8, 5, 7$ 16. $-9, -6, -3$ 17. $-2, -1, 0, 1$
18. $-2, -1, 5, 7$ 19. $-23, -20, -14, 14$
20. $-3, -2, -1, 0, 1, 2, 3$ 21. $-5 < 4$ and $4 > -5$
22. 1994; $-\$5,000,000 > -\$9,000,000$
23. Possible answer: $101 > 88$. The Bulls scored 101 points. The Celtics scored 88 points. Which team won?

4·5 Adding Integers

1. 2 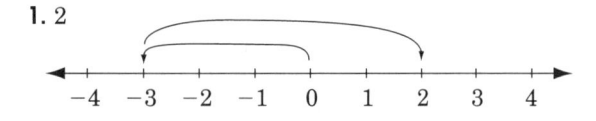
2. $-2 + 6 = 4$
3. $-1 + (-3) = -4$
4. $1 + (-2) = -1$
5. -4 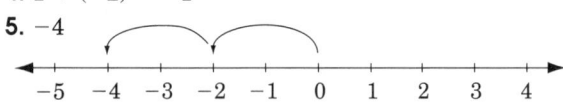
6. 3 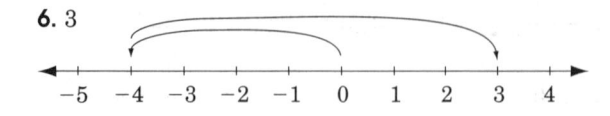
7. 2 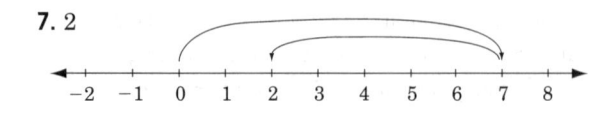
8. -8 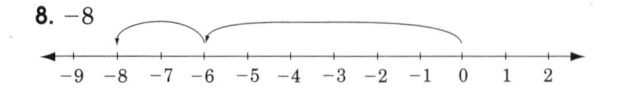
9. -2 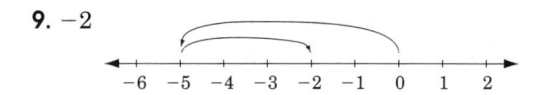
10. No. Simon should have moved 8 units to the left.
11. Add 12. Subtract 13. Subtract 14. Add
15. 1 16. -12 17. -1 18. 0 19. -15 20. 29
21. -4 22. -24 23. 7 24. 3 25. -12 26. 39
27. -1 28. -3
29. Possible answer: Wally the Worm was 8 feet below the ground. He climbed up 5 feet. Where is Wally now?

4·6 Subtracting Integers

1. $6 + (-10)$ 2. $-4 + 3$ 3. $8 + 2$ 4. $-4 + (-3)$
5. -3 6. 11 7. -11 8. 15 9. -15 10. -39
11. -25 12. 72 13. -100 14. 0 15. 3 16. 13
17. -2 18. -12 19. Profit; $1,450
20. Possible answer: Susan gained 6 pounds one month, then lost 10 pounds the next month. What is Susan's total change in her weight?

4·7 Multiplying Integers

1. Negative 2. Positive 3. Positive 4. Negative
5. -18 6. 16 7. 63 8. 0 9. 80 10. 84 11. -480
12. -15.66 13. -8 14. $\frac{15}{16}$ 15. -120 16. -540
17. -4 and -5 18. -980 ft
19. Possible answer: If the number of factors is even, the product is positive. If the number of factors is odd, the product is negative.

4·8 Dividing Integers

1. Negative 2. Positive 3. Negative 4. Positive
5. -5 6. -5 7. 2 8. -7 9. 3 10. -4 11. 6
12. 7 13. -5 14. -12 15. 14 16. -7 17. 3
18. 18 19. -30 20. -5 21. a. $-18°F$ b. $-3°F$
22. a. $6°C$ b. $2°C$ c. 2 hours
23. Marsha. Possible answer: Karen is wrong because 0 can be divided by 8. Ted is wrong because $64 \div 8 = 8$, not $0 \div 8$.

Chapter 4 Review

1. g 2. b 3. h 4. a 5. d 6. c 7. f 8. e 9. 23
10. 0 11. -3 12. 2 13. 5 14. 9 15. 17.8 16. 7
17–19. 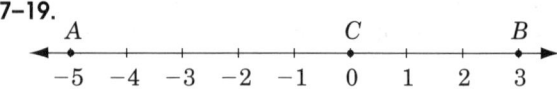
20. $-6, -3, 3, 6$ 21. $-8, -5, -2, -1, 0, 1$ 22. 2
23. -17 24. 7 25. -72 26. -11 27. 17 28. 400

29. −7 **30.** −4 **31.** −5 **32.** = **33.** < **34.** > **35.** >
36. −12 ft **37.** −20¢
38. Answers will vary. Student game should include negative and positive numbers.

Chapter 4 Practice Test

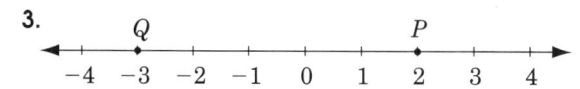

1. −15 **2.** 2
3.

```
         Q                    P
  ←——————•————————————————•———————→
     -4 -3 -2 -1  0  1  2  3  4
```

4. 13 **5.** 7 **6.** 14 **7.** 10 **8.** 14 **9.** 0
10. −3, −1, 0, 4 **11.** −5, −1, 0, 1, 5 **12.** 9 **13.** 28
14. −1 **15.** 3 **16.** 5 **17.** −40 **18.** 3 **19.** −28
20. −64 **21.** −5 **22.** −9 **23.** −5 **24.** > **25.** <

Chapter 4 Test

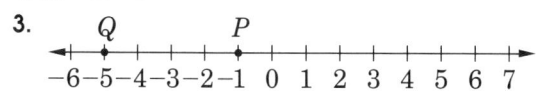

1. 13 **2.** −1
3.

```
         Q        P
  ←——•———•——•——•—•——•——•——•——•——•——•——•——•—→
    -6 -5 -4 -3 -2 -1  0  1  2  3  4  5  6  7
```

4. 7 **5.** 9 **6.** 16 **7.** 17 **8.** 17 **9.** 11
10. −4, −1, 0, 1, 3 **11.** −4, −2, 1, 2, 4
12. 11 **13.** 30 **14.** −4 **15.** −7 **16.** 7
17. −28 **18.** 9 **19.** −39 **20.** −70 **21.** 1
22. 4 **23.** 12 **24.** = **25.** <

Chapter 5 Introduction to Graphing

5•1 The Coordinate Plane

1. 2 west, 4 south **2.** 5 west, 2 north
3. 3 west, 0 north or south
4. 0 east or west, 0 north or south
5–8.

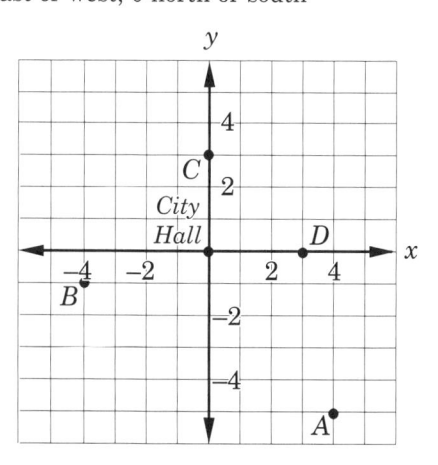

9. NW **10.** SE **11.** NE **12.** SW
13. 9 blocks **14.** 45 blocks
15. Possible answer: A coordinate plane has two directions, horizontal (east and west) and vertical (north and south). A coordinate shows the movement in only one direction, so two are needed.

5•2 Ordered Pairs

1. (4, 1) **2.** (−2, 0) **3.** (3, −4) **4.** (0, −2) **5.** (−3, 4)
6–11.

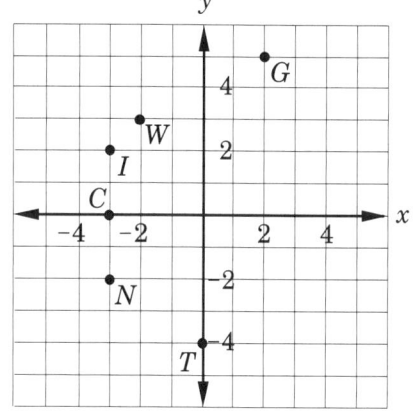

12. II **13.** IV **14.** III **15.** 4 **16.** 5 **17.** (6, −5)
18. No. Possible answer: The order of the coordinates is important in an ordered pair. (2, 5) is 2 to the right and 5 up, while (5, 2) is 5 to the right and 2 up.

5•3 Tables and Graphs

1. 7
2. 13
3. −2
4. 4
5.

x	y	(x, y)
−2	0	(−2, 0)
0	2	(0, 2)
1	3	(1, 3)

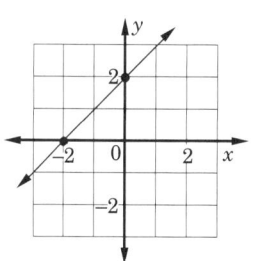

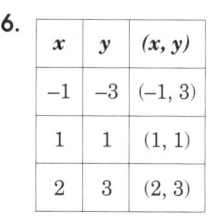

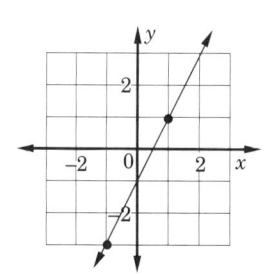

6.

x	y	(x, y)
−1	−3	(−1, 3)
1	1	(1, 1)
2	3	(2, 3)

Answers **45**

7. a.

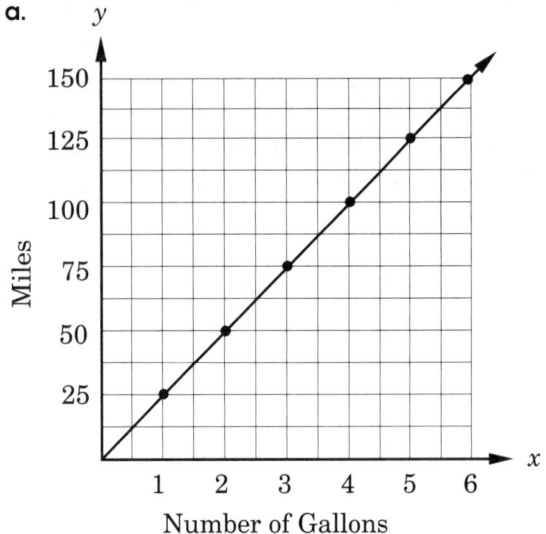

(Graph: Miles vs. Number of Gallons, linear line through (1,25), (2,50), (3,75), (4,100), (5,125), (6,150))

b. Possible answer: The car gets 25 miles per gallon.

8. Possible answer: 5. Find $-1\frac{1}{2}$ on the y-axis, then go across to line. Find the x-value for the point.

5•4 Special Graphs

1. Continuous 2. Horizontal and continuous
3. Discrete 4. Vertical and continuous 5. $x = -3$
6. $y = -7$ 7. Discrete 8. Continuous
9. Yes. Possible answer: A vertical line and a horizontal line will intersect at one point. These lines intersect at (5, –3).

5•5 Slope

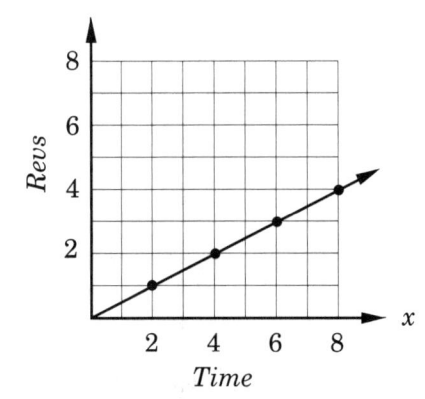

1. $\frac{3}{2}$ 2. $(-3, 4); (1, 1); -\frac{3}{4}$ 3. $(1, -2); (4, 1); 1$
4. $(-4, 2); (-1, -3); -\frac{5}{3}$ 5. $\frac{1}{2}$ 6. $\frac{4}{3}$ 7. 9 8. -1
9. 3 10. -5 11. $-\frac{4}{7}$ 12. 0 13. $\frac{1}{2}$

(Graph: Revs vs. Time, line through points (2,1), (4,2), (6,3), (8,4))

14. Possible answer: Lines with a positive slope extend from the lower left to the upper right; lines with a negative slope extend from the upper left to the lower right.

5•6 Finding the Intercepts of an Equation

1. 5; 3 2. –2; 1 3. –4; –1 4. –4; 4 5. 3; –6
6. –6; 3 7. $\frac{8}{5}$; –8 8. 3; –4 9. 5; 7 10. (–8, 0)
11. –40
12. Answers will vary, but should be in the form $y = k$, with k not equal to zero.
13. **a.** 32°F **b.** 40°C
    c. slope: $\frac{9}{5}$; y-intercept: 32
14. Possible answer: Since two points determine a line, finding the x- and y-intercepts would give you the two necessary points.

Chapter 5 Review

1. axes
2. continuous line
3. coordinate plane
4. coordinates
5. graph of an equation
6. linear equation
7. origin
8. 14 9. 15
10. (–3, –4) 11. (3, 2) 12. (0, –1)
13–15.

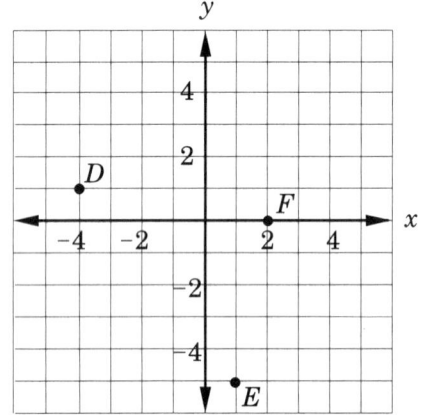

(Coordinate plane graph with points D, E, F)

16. IV 17. III 18. 7 19. 25 20. –1 21. 18.5
22.

x	y	(x, y)
-1	-4	(-1, -4)
1	0	(1, 0)
3	4	(3, 4)

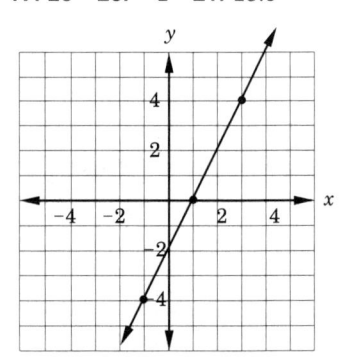

(Graph of line through these points)

23. $x = 6$ 24. –3 25. 2 26. $\frac{1}{4}$ 27. 6; –6 28. 7; –4
29. 12 ft 30. Answers will vary.

Chapter 5 Practice Test

1. $(-3, -2)$ 2. $(3, 0)$

3 and 4

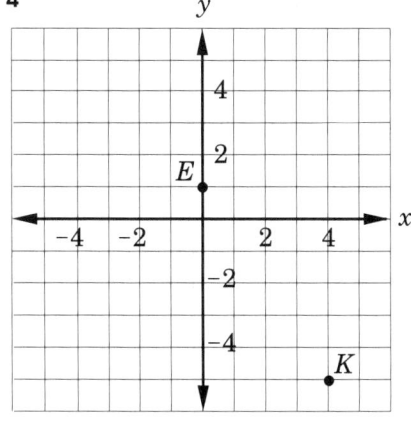

5. I   6. IV   7. 12   8. 20   9. -11 10. 14

11.

x	y	(x, y)
-2	5	(-2, 5)
0	3	(0, 3)
2	1	(2, 1)

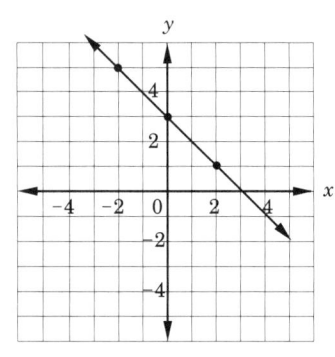

12. $y = 9$ 13. $\frac{2}{5}$ 14. $\frac{1}{2}$ 15. $\frac{5}{8}$ 16. $2; 8$ 17. $5; -3$
18. $1; 1$ 19. $0; 0$ 20. $-1; 2$

Chapter 5 Test

1. $(-1, -2)$ 2. $(4, -3)$

3. & 4.

5. II   6. I   7. 9   8. 10   9. 7   10. -24.5

11.

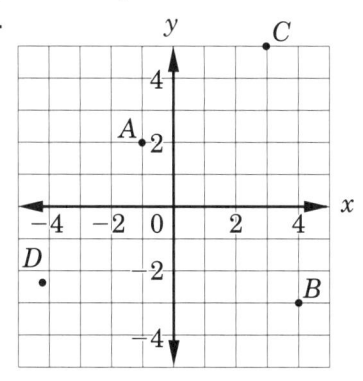

x	y	(x, y)
2	-1	(2, -1)
3	1	(3, 1)
0	-5	(0, -5)

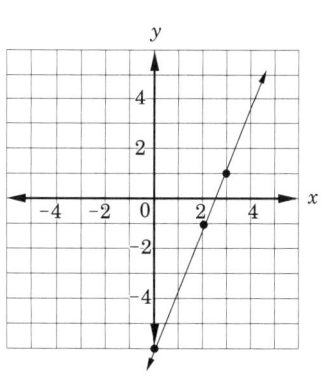

12. $x = 6$ 13. $\frac{9}{7}$ 14. $\frac{2}{5}$ 15. -3 16. $-\frac{5}{3}; 5$
17. $2; -4$ 18. $2; 16$ 19. $0; 0$ 20. none; 8

Chapter 6 More About Graphing

6•1 Slope as Rate of Change

1. 0.03   2. 2   3. -3 4. $\frac{4}{7}$ 5. $+9¢$
6. Possible answer: Multiply the yearly rate of change by 10.

6•2 Using the Slope-Intercept Form of an Equation

1. Slope: -2; y-intercept: -5
2. Slope: 3; y-intercept: $-\frac{1}{2}$
3. Slope: $-\frac{2}{3}$; y-intercept: 4   4. $y = 2x + 5$
5. $y = 4x - 1$ 6. $y = -\frac{5}{2}x + 12$ 7. $y = \frac{2}{3}x - 1$
8. $\frac{1}{8}x + 6 = y$
9. Possible answer: The line with slope $\frac{1}{5}$ will only extend up 1 unit for every 5 units across. The slope of 5 is much steeper: the line extends up 5 units for every 1 unit across.

6•3 Other Forms of Equations

1. $-5x + 3y = 10$ 2. $4x - y = 3$ 3. $4x - 9y = 7$
4. $2x + 6y = 3$ 5. $4x - 10y = -2$
6. $12x - 15y = 10$ 7. $y = \frac{1}{2}x + 2$; student graph
8. $y = -x - 3$; student graph   9. $48 = 6x + 8y$
10. Possible answers: Jim bought some walnuts for $3.50 a pound and cashews for $4.00 a pound. If he spent $10.00, how much of each kind of nut did he buy? Possible equation: $3.5x + 4.00y = 10.00$

Answers 47

6·4 Graphing Absolute-Value Equations

1. $y = 8$ 2. $y = 3$ 3. $y = 3$ 4. $y = 6$
5.
$x = -6$	$y = 2$
$x = -5$	$y = 1$
$x = -4$	$y = 0$
$x = -3$	$y = 1$
$x = -2$	$y = 2$

6. Possible answers: $x = -2, -1, 0, 1, 2$
   $y = 2, 1, 0, 1, 2$; student graph
7. Possible answers: $x = -2, -1, 0, 1, 2$
   $y = 3, 2\frac{1}{2}, 2, 1\frac{1}{2}, 1$; student graph
8. c
9. No; Possible answer: The graph of $y = 2x$ will include negative values for x and will be in the first and third quadrants. The graph of $y = |2x|$ will have only positive values for y and will be in the first and second quadrants.

6·5 Other Graphs and Their Equations

1. perpendicular. 2. parallel 3. perpendicular
4. neither 5. 4 6. 5 7. 6 8. $-\frac{5}{3}$
9.
$x = 0$	$y = 0$
$x = 2$	$y = 6.4$
$x = 4$	$y = 9.6$
$x = 5$	$y = 10$
$x = 6$	$y = 9.6$
$x = 8$	$y = 6.4$
$x = 10$	$y = 0$; student graph

10. Possible answer: The slope of a line perpendicular to a given line is the negative reciprocal of the slope of the given line.

Chapter 6 Review

1. perpendicular 2. quadratic 3. y-intercept form
4. $\frac{3}{-3} = -1$ 5. $\frac{-8}{-2} = 4$ 6. $y = 3x - 2$
7. $y = -\frac{3}{4}x + 4$ 8. Slope: 1; y-intercept: 1
9. Slope: 4; y-intercept: 2 10. Slope: -2; y-intercept: 5
11. Slope: $\frac{1}{2}$; y-intercept: $\frac{3}{2}$ 12. $4x - 3y = 1$
13. $2x + 6y = 8$ 14. -5 15. $\frac{-5}{2}$
16. Possible answers: $x = -2, -1, 0, 1, 2$
    $y = 3, 2, 1, 0, 1$
17. Possible answers: $x = -2, -1, 0, 1, 2$
    $y = -2, -5, -6, -5, -2$
18. $\frac{-377}{23}$ 19. $\frac{12}{5}$
20. Answers will vary but should show that students observed that the ball rose, hit a high point, and then fell. Their data should indicate a pattern similar to a parabola.

Chapter 6 Practice Test

1. $\frac{1}{2}$ 2. $y = -2x - 6$ 3. Slope: -4; y-intercept: 6
4. $y = \frac{4}{5}x + 1$ 5. $4x - 5y = -5$ 6. 1 7. $\frac{3}{2}$
8. $1.51 per year
9. Possible ordered pairs:

x	-4	-3	-2	-1	0
y	0	1	2	3	4

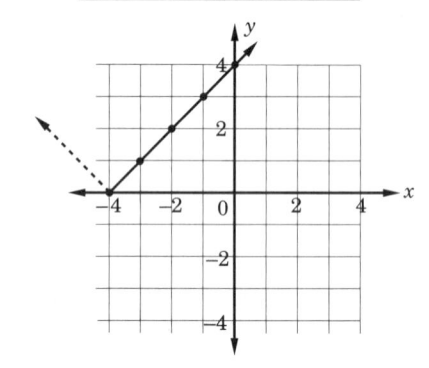

10.
x	-2	-1	0	1	2
y	0	3	4	3	0

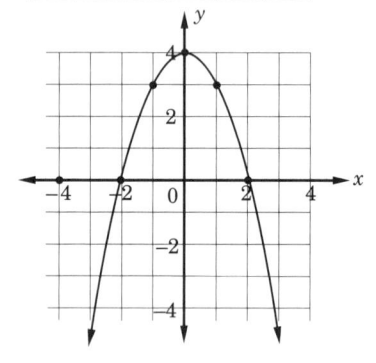

Chapter 6 Test

1. $-\frac{3}{5}$ 2. $y = 3x - 7$ 3. Slope: 4 y-intercept: -12
4. $y = \frac{2}{3}x + 3$
5. $8x - 12y = -36$ or $2x - 3y = -9$
6. $-\frac{4}{3}$ 7. $\frac{2}{3}$
8. $\frac{16}{7}$ or $2.29 per year
9.
x	0	1	2	3	4
y	3	2	1	0	1

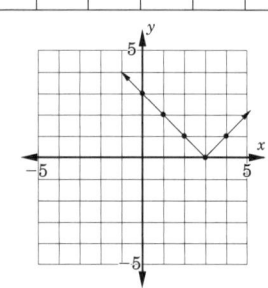

48 Answers

10.

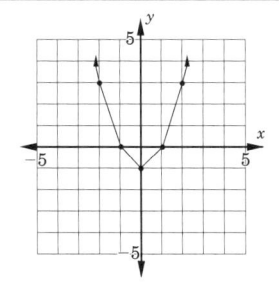

x	-2	-1	0	1	2
y	3	0	-1	0	3

Chapter 7 Exponents

7•1 More About Exponents

1. 6^4 2. 4^5 3. 2^7 4. $(13.7)^3$ 5. $3 \cdot 3 \cdot 3 \cdot 3 \cdot 3$
6. $20 \cdot 20 \cdot 20 \cdot 20$ 7. $9 \cdot 9$ 8. $11 \cdot 11 \cdot 11$
9. $x \cdot x \cdot x \cdot x \cdot x$ 10. $n \cdot n$ 11. 11 12. 10,000
13. 125 14. 81 15. 64 16. 729 17. 2,097,152
18. 512
19. No. $3^2 = 9$ and $2^3 = 8$, so they are not equal.
20. 1,808.64 cm², 7,240.32 cm³
21. Possible answer: Squaring a number means multiplying it by itself, whereas multiplying a number by 2 merely doubles the number.

7•2 Writing Very Large and Very Small Numbers

1. No; 45.6 is not between 1 and 10.
2. Yes 3. No 4. No 5. Yes 6. Yes 7. No
8. 4; 10^4 9. 6; 10^6 10. 8; 10^8 11. $7 \cdot 10^6$
12. $4.36 \cdot 10^3$ 13. $5.307 \cdot 10^4$ 14. $9.031 \cdot 10^7$
15. $8.8008 \cdot 10^{10}$ 16. $5.047 \cdot 10^{10}$ 17. 935,600
18. 14,400,000 19. 3,070 20. 62,040
21. $3.4 \cdot 10^8$ The number will be larger if the exponent is positive rather than negative.
22. $9.6 \cdot 10^{-2}$ 23. $9.32 \cdot 10^{-4}$ 24. $7.05 \cdot 10^{-5}$
25. $5.810 \cdot 10^{-6}$ 26. $4.7 \cdot 10^{-7}$ 27. $5.206 \cdot 10^{-3}$
28. 0.0000138 29. 0.007064 30. 0.0803
31. 0.000001005 32. $5 \cdot 10^{-3}$ 33. $\$4.09 \cdot 10^{50}$
34. 93,000,000 miles
35. Possible answers: Astronomers use very large numbers to measure distances to planets and the speeds of light and sound. Microbiologists use very small numbers to measure the width of atoms.

7•3 Multiplication Properties of Exponents

1. 6^7 2. 6^{10} 3. 9^7 4. 9^{12} 5. 2^{21} 6. 2^{10}
7. $125k^3$ 8. $27p^3$ 9. a^{11} 10. m^{11} 11. x^4y^6
12. $64n^8$ 13. a. 2^4p b. 16
14. No. $5^3 \cdot 5^4 = 5^7$, but $(5^3)^4 = 5^{12}$.

7•4 Division Properties of Exponents

1. 3^5 2. $\frac{8}{27}$ 3. 5^2 or 25 4. 9 5. 10^4 or 10,000
6. 4^{10} 7. $\frac{1}{32}$ 8. $\frac{81}{10,000}$ 9. $\frac{a^3}{64}$ 10. $\frac{m^9}{n^9}$ 11. $2x^4$
12. $5p^8$ 13. 2^2, or 4
14. A little over 100 times bigger
15. Possible answer: A rocket can travel 4,650 miles per hour. The sun is about 93,000,000 miles from earth. How long will it take for the rocket to reach the sun? Write $\frac{93,000,000}{4,650}$ as $\frac{9.3 \times 10^7}{4.65 \times 10^3}$, which equals $2 \cdot 10^4$ or 20,000 hours for the rocket to reach the sun.

7•5 Negative Exponents

1. $\frac{1}{8}$ 2. $\frac{1}{64}$ 3. $\frac{1}{25}$ 4. $\frac{1}{81}$ 5. $\frac{1}{128}$ 6. $\frac{1}{100,000}$ 7. 1
8. $\frac{1}{9}$ 9. $\frac{1}{144}$ 10. $3^{-3} = \frac{1}{3^3}$ 11. $2^{-8} = \frac{1}{2^8}$ 12. $n^{-1} = \frac{1}{n}$
13. $k^{-5} = \frac{1}{k^5}$ 14. $\frac{7}{y^3}$ 15. $b^{-5} = \frac{1}{b^5}$ 16. $5p^9$ 17. -2
18. $c^{-4} = \frac{1}{c^4}$ 19. $\frac{1}{12z^4}$ 20. $\frac{1}{8k^6}$ 21. $3x^7y^2$ 22. $P = \frac{5F}{4D^2}$
23. Possible answer: Agree. Whatever the exponent, 1 will be the only factor, so the result will be 1.

7•6 The Exponent of Zero

1. 1 2. 1 3. 1 4. 1 5. 1 6. 1 7. $5^0 = 1$
8. $m^0 = 1$ 9. 3 10. -15 11. $4a$ 12. $5c^{-4} = \frac{5}{c^4}$
13. $8h^4k^3$ 14. $32n^4$ 15. 8, 4, 2, 1, $\frac{1}{2}, \frac{1}{4}, \frac{1}{8}, \frac{1}{16}$ 16. -7
17. $\frac{1}{3}$ or $-\frac{1}{3}$ 18. 0
19. Agree. Possible answer: $2^{-2} = \frac{1}{4}$, $2^0 = 1$, and $2^2 = 4$, so it seems that $2^x > 0$ for any value of x.

7•7 Radicals

1. Yes. Possible answer: Since x^2 cubed is $(x^2)^3 = x^6$, the cube root of x^6 is x^2.
2. 7 3. $5x^3$ 4. 9 5. -4 6. -1 7. 10 8. 2 9. 1
10. $8h$ 11. $-\frac{7}{10}$ 12. True 13. True
14. 45.9684; Possible answer: 6.78 is only an approximation of $\sqrt{46}$, not an exact value.
15. 3.2 16. 7.8 17. 1.4 18. 4.8 19. 9.2 20. 2.6
21. 4.24 22. 6.32 23. 10 in. 24. 8 ft
25. 127.28 ft 26. 154.92 ft 27. 47.4 sec
28. Possible answer: The calculator will display an error message, because it is not possible to take the square root of a negative number.

Chapter 7 Review

1. square 2. perfect square 3. radical

Answers 49

4. power or exponent 5. 9^4 6. b^3 7. 32
8. 100,000 9. $6.45 \cdot 10^4$ 10. $9 \cdot 10^{-4}$
11. $3.31 \cdot 10^{-7}$ 12. $1 \cdot 10^{11}$ 13. 0.05 14. 999
15. 0.0000010302 16. 81,100,000 17. 7^8 18. k^{10}
19. $27n^3$ 20. $2^{20}p^{18}$ 21. $\frac{1}{8}$ 22. $\frac{1}{64}$ 23. $\frac{1}{5}$ 24. $\frac{c^4}{7^4}$
25. $2x^6$ 26. 27 27. 9 28. $\frac{\sqrt{3}}{8}$ 29. $-7x^4$ 30. $\frac{4}{5}$
31. 1.1 sec 32. 50 mi/h
33. Possible answers: Numbers might represent total sales of a company, population of a country, or the number of people attending a concert.

Chapter 7 Practice Test

1. 3^6 2. x^4 3. 125 4. 256 5. 1 6. $\frac{1}{36}$
7. $6.7 \cdot 10^{-4}$ 8. $4.56 \cdot 10^8$ 9. 2,000 10. 0.00008001
11. 8^9 12. n^{10} 13. $81a^4$ 14. $\frac{1}{5^5}$, or $\frac{1}{3,125}$ 15. $\frac{1}{x^2}$
16. $\frac{c^5}{32}$ 17. $\frac{3}{y^3}$ 18. x^4 19. 2 20. 63.2 mi/h

Chapter 7 Test

1. 4^5 2. y^6 3. 64 4. 729 5. 1 6. $\frac{1}{125}$
7. $5.83 \cdot 10^{11}$ 8. $1.35 \cdot 10^{-4}$ 9. 51,000
10. 0.00763 11. 3^{10} 12. x^6 13. $8b^3$ 14. $\frac{1}{6^6}$ 15. x^6
16. $\frac{x^4}{81}$ 17. $3x^3$ 18. y^3 19. 3 20. 24.5 ft

Chapter 8 Inequalities

8•1 Equations and Inequalities

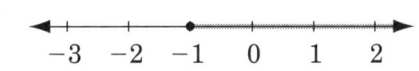

1. $x < 4$ 2. $x \geq -5$ 3. $x \leq -1$ 4. $x > 3$
5.

   ———•⟶
   $-3\ -2\ -1\ 0\ 1\ 2$

6.

   ⟵∘———
   $-4\ -3\ -2\ -1\ 0\ 1$

7.

   ———•⟶
   $-1\ 0\ 1\ 2\ 3\ 4\ 5$

8.

   ⟵∘———
   $-2\ -1\ 0\ 1\ 2$

9. $x < 5$ 10. $x > 4$ 11. $x \leq 3$ 12. $x \geq 4$
13. $4,600
14. Possible answer: "No more than" means equal to or less than, so you would use the symbol $\leq$.

8•2 Solving One-Step Inequalities

1. Should be reversed 2. Stays the same
3. Stays the same 4. Should be reversed 5. $x < -6$
6. $x \geq 34$ 7. $x < -5$ 8. $x \geq 5$ 9. $n > 52$
10. $x \geq -6$ 11. a. $l - 0.03 \leq l \leq l + 0.03$ b. 10.7 cm
12. $c > \$30$ and $c < \$250$
13. You have less than 100 sticks of licorice. You want to give 20 people an equal number of sticks. How could you decide how many sticks each person gets?

8•3 Solving Two-Step Inequalities

1. Subtract 4. 2. Add 6. 3. Subtract 5. 4. $x < 8$
5. $x \geq 4$ 6. $x < -4$ 7. $x \geq 78$ 8. $x < 2$ 9. $x > 48$
10. $x < -2$ 11. $x \geq 7$ 12. $x \leq -4$
13. a. $300 + 40h \leq 1,600$ b. $32.5\ h$
14. $3r + 550 \leq 1,600$
    $3r \leq 1,050$
    $r \leq 350$

   ⟵———————•———⟶
   0 50 100 150 200 250 300 350 400

15. Possible answer: $-2x - 5 \geq 15$ or $\frac{x}{2} + 5 \leq 0$

Chapter 8 Review

1. inequality 2. Tolerance 3. budget 4. $x \geq 2$
5. $x < 5$ 6. $x > -2$ 7. $x \leq -3$
8.

   ———∘———⟶
   $-1\ 0\ 1\ 2\ 3\ 4\ 5$

9.

   ⟵———•———
   $-2\ -1\ 0\ 1\ 2$

10.

    ⟵∘———
    $-4\ -3\ -2\ -1\ 0\ 1$

11.

    ———•————
    $-6\ -5\ -6\ -3\ -2\ -1\ 0$

12. Stays the same 13. Stays the same
14. Should be reversed 15. Should be reversed
16. Divide by 5. 17. Multiply by -6. 18. Add 6.
19. Subtract 14. 20. $x > 4$ 21. $p \leq 17$ 22. $k > -20$
23. $x \geq -15$ 24. $c < 6$ 25. $x \leq -27$ 26. $x > 10$
27. $p > 3$ 28. $h \leq 48$
29. a. $n \leq 50$, $n > 50$, and $n \leq 100$ b. $457.50
30. Answers will vary, but should include expressions of inequality such as, at most, less than, more than, at least, etc.

Chapter 8 Practice Test

1. $x \geq -3$ 2. $x < -1$

3.

$\leftarrow\!\!+\!\!\!+\!\!\!+\!\!\!+\!\!\!+\!\!\!+\!\!\!+\!\!\!\bullet\!\!\!+\!\!\!\rightarrow$
-101234

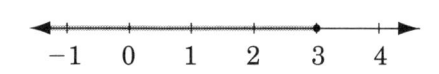

4.

$\leftarrow\!\!+\!\!\!+\!\!\!\circ\!\!\!+\!\!\!+\!\!\!+\!\!\!+\!\!\!+\!\!\!+\!\!\!\rightarrow$
$-3-2-1012$

5. Should be reversed **6.** Stays the same
7. Stays the same **8.** Stays the same
9. Stays the same **10.** $x \geq 5$ **11.** $x < 14$
12. $k > -5$ **13.** $p \geq 5$ **14.** $n > 35$ **15.** $x > 3.5$
16. $n < -5$ **17.** $x \geq 3$ **18.** $x > 8$
19. $800 + 0.04s \geq 3{,}000$
$s \geq 55{,}000$
20. $6n + 3 \geq 63$
$n \geq 10$

Chapter 8 Test

1. $x > -1$ **2.** $x \leq 2$
3.

$\leftarrow\!\!+\!\!\!+\!\!\!+\!\!\!+\!\!\!\bullet\!\!\!+\!\!\!+\!\!\!\rightarrow$
$-2-1012$

4.

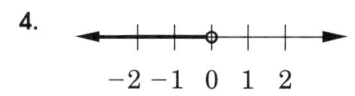

$\leftarrow\!\!+\!\!\!+\!\!\!+\!\!\!\circ\!\!\!+\!\!\!+\!\!\!\rightarrow$
$-2-1012$

5. Stays the same **6.** Stays the same
7. Should be reversed **8.** Should be reversed
9. Stays the same **10.** $x \leq 7$ **11.** $k \leq -5$
12. $25 < t$ **13.** $m \geq -18$ **14.** $-2.9 < x$
15. $n > -1$ **16.** $1 < x$ **17.** $x \geq -10$
18. $x > 18$ **19.** $x > \$7.50$ **20.** $x \geq 3\frac{1}{3}\%$

Chapter 9 Monomials and Polynomials

9•1 Recognizing Monomials and Polynomials

1. trinomial **2.** monomial **3.** binomial **4.** binomial
5. 4 **6.** 5 **7.** 3 **8.** 1 **9.** 4 **10.** 4 **11.** 5 **12.** 3
13. $3b, 7b, 4b$ **14.** $x, 3x, 9x$ **15.** $2ac^2, 9ac^2$
16. a. s^3 **b.** m^3 **c.** $(2d)^3 = 8d^3$
17. Possible answer: Each variable is to the first degree. In the product lwh you add exponents to find the degree of 3.

9•2 Adding and Subtracting Polynomials

1. $5n + 1$ **2.** $3a^2 + a - 1$ **3.** $22ab$ **4.** -6
5. $2x^2 - 2x + 16$ **6.** $14x^2 + 3x - 4$
7. $-x^2 + 6x - 9$ **8.** $6x$

9. $2x^3 + 5y^2 + 27$ cubic units
10. Possible answer: Add 5.3 ft + 9 inches. You cannot add unless you convert 5.3 ft to inches or inches to feet. 5.3 ft = 63.6 inches, so 5.3 ft + 9 inches = 72.6 inches. Alternatively, 5.3 ft + 9 inches = 5.3 ft + 0.75 ft = 6.05 ft.
You can only add like dimensions. However, in measurements you can convert from one linear dimension to another, or one square area dimension to another using the correct conversion factor. For polynomials, you cannot convert from x to y or from x^2 to y^2 using any conversion factor.

9•3 Multiplying a Polynomial by a Monomial

1. $27rw$ **2.** $12a^2 + 15a^2b$ **3.** $14x^3$ **4.** $15n^2 + 40n$
5. $24a^2b + 20ab^2$ **6.** $6x^2 + 9x + 18$
7. $40x^3 - 40x^2 - 64x$ **8.** $n^6 + n^5 + n^4 + n^3 + n^2$
9. $7a^3 + 56a^2$ **10.** $2x^3 + 14x^2$
11. $8x^3 + 12x^2 - 24x$
12. $-21v^4 + 14v^3 + 49v^2 + 63v$
13. $-9st^2 - 9st + 81$
14. $6a^3cd + 18a^2c^2d - 24acd^2$
15. $12k^2 - 12k + 36$
16. $4m^6 - 20m^4 + 14m^3$
17. Jacob's intermediate step was correct as $30x^5 + 42x^4 + 12x^3$. This is the final answer. Jacob's mistake is adding unlike terms together. You cannot add terms in x^5, x^4, x^3 together to get a single term in x^{12}.
18. $-2x^5 - 2x^4 - 5x^3$
19. $10a^3bc + 15ab^3c + 20abc^3$
20. $4a^3b^3c^4 - 4a^2b^4c^4 - 2a^2b^3c^5$
21. $-2x^3y^3z^2 - 2x^3y^3z^2 - 2x^3y^3z^2 = -2x^3y^3z^2 - 4x^3y^3z^2$
22. $64r^4s^2t^3 + 24r^2s^3t^2 + 32r^2s^2t^2 + 64rst$
23. $92(w^2 + 15w)$
24. $C = 92(w^2 + 15w)$; $185,288
25. Possible answer: Use exercise 18.
$-x^3(2x^2 + 2x + 5)$ Distributive
$= (-x^3 \cdot 2x^2) + (-x^3 \cdot 2x) + (-x^3 \cdot 5)$ Commutative
$= (-2x^2 \cdot x^3) + (-2x^3 \cdot x) + (-5x^3)$ Associative
$= (-2)(x^2 \cdot x^3) + (-2)(x^3 \cdot x) + (-5)(x^3)$
 Multiplicative
$= -2x^5 - 2x^4 - 5x^3$

9•4 Multiplying Polynomials

1. $n^3 - n^2 + n - 6$
2. $n^2 + 5n + 6$
3. $n^2 + n - 20$
4. $n^3 + 7n^2 + 17n + 35$
5. $10x^2 + x - 24$

6. $x^3 + x^2 - 12$
7. $2n^3 - 13n^2 + 22n - 8$
8. $xy + 3y - 5x - 15$
9. $3x^3 - 13x^2 - 46x - 24$
10. $6n^4 + 19n^3 + 30n^2 + 13n - 20$
11. $2x^4 + 5x^3 - 13x^2 + 19x - 28$
12. He is correct. He has not simplified by adding like terms. The final answer is $x^2 + x - 6$.
13. $x^2 + 16x + 63$
14. $x^2 - 11x + 30$
15. $x^2 - 7x - 18$
16. $x^2 + x - 56$
17. $2x^2 + 7x + 6$
18. $6x^2 - 23x + 21$
19. $28x^2 + 13x - 63$
20. $6x^2 - 17x - 45$
21. $(3x + 4)(x + 4) = 3x^2 + 16x + 16$ in.^{2}
22. Possible answer: If you multiply a binomial times a polynomial, you get more than 4 terms. You might have a binomial times a trinomial which would give 6 terms, not 4, to be added.

Chapter 9 Review

1. c 2. e 3. d 4. a 5. b 6. trinomial; 3
7. monomial; 9 8. binomial; 1 9. trinomial; 2
10. $11j - 2$ 11. $3x^2 - 9$ 12. $7w + 8$
13. $x^3 + 3x^2 + 12x$ 14. $2a^2 + 3ac$ 15. $-2d + 2$
16. $32bc$ 17. $4n^3 + 12n^2 - 28n$
18. $6a^2b^2c + 8a^2b^3c + 10ab^2c^2$
19. $x^2 + 12x + 32$ 20. $6x^2 - 13x - 28$
21. $4x^4 + 12x^3 + x^2 - 22x - 15$
22. Each shelf area = 18 in. • 36 in. = 648 in.^{2}. Each cupboard has 2 shelves and she has 6 cupboards. Total shelf area = 6 • 2 • 648 in.^{2} = 7,776 in.^{2}. One roll of shelf paper covers 1,000 in.^{2}. Tina must buy 8 rolls of shelf paper.
23. 269 in.^{2}
24. 165,248 in.^{3}; 165,248 peanuts
25. Possible answer: Sales tax 7.75%: let x = cost of item. Then: purchase price = $x + 0.0775x$ = $1.0775x$.

Chapter 9 Practice Test

1. Binomial, 1 2. Monomial, 9 3. Trinomial, 3
4. Trinomial, 2 5. $4h + 1$ 6. $3x^2 - 23$
7. $-3x^3 + 6x^2 + 17x$ 8. $x^3 + 9x^2 - 7x$ 9. $-2a - 3$
10. $-3k$ 11. $3n^2 - 8n$ 12. $7x^5y^2$ 13. $18x^2 + 81x + 81$
14. $6x^3 + 18x^2 - 21x$ 15. $b^2 + 14b + 48$
16. $12x^2 - 44x + 35$
17. $2x^4 - 3x^3 - 5x^2 - 7x - 15$
18. $3x^4 - 2x^3 - 32x^2 - 68x - 48$

19. $x(3x)(x + 4) = 3x^3 + 12x^2$
20. $2x^3 - (\frac{1}{2}x)^3 = 2x^3 - \frac{1}{8}x^3 = \frac{15}{8}x^3$ cubic units of foam "peanuts."

Chapter 9 Test

1. Binomial; 3 2. Trinomial; 2 3. Monomial; 7
4. Trinomial; 4 5. $3x^2 - 4x - 5$ 6. $3x^2 + 4x + 6$
7. $12a^2 - 4b$ 8. $32x^2 - 40x + 16$ 9. $x^2 + 9x - 7$
10. $18x^3 - 24x^2 + 12x$ 11. $4x^7y^3$
12. $20x^2 + 11x - 3$ 13. $30x^3y^3 - 24x^2y^2 + 36xy$
14. $32x^4 - 56x^3 + 72x^2 - 24x$
15. $25x^2 + 20x + 4$ 16. $-6x^2 - x + 1$
17. $15a^2bc - 30ab^2c$ 18. $4x^2 - 4x - 3$
19. $30x^3 + 103x^2 + 76x + 15$ 20. $25x^2 + 50x$

Chapter 10 Factoring

10•1 Finding the Greatest Common Factor

1. Yes; 7 divides evenly into 84.
2. The GCF is 5.
3. 45: 1, 3, 5, 9, 15, 45; 65: 1, 5, 13, 65; GCF is 5
4. 15: 1, 3, 5, 15; 24: 1, 2, 3, 4, 6, 8, 12, 24; GCF is 3
5. 80: 1, 2, 4, 5, 8, 10, 16, 20, 40, 80; 104: 1, 2, 4, 8, 13, 26, 52, 104; GCF is 8
6. 12: 1, 2, 3, 4, 6, 12; 30: 1, 2, 3, 5, 6, 10, 15, 30; 42: 1, 2, 3, 6, 7, 14, 21, 42; GCF is 6
7. 18: 1, 2, 3, 6, 9, 18; 36: 1, 2, 3, 4, 6, 9, 12, 18, 36; 72: 1, 2, 3, 4, 6, 9, 12, 18, 36, 72; GCF is 18
8. 13: 1, 13; 18: 1, 2, 3, 6, 9, 18; GCF is 1
9. 8: 1, 2, 4, 8; 15: 1, 3, 5, 15; 49: 1, 7, 49; GCF is 1
10. Yes. Possible answer: Because there are no more factors in common after $5uv^2$.
11. $6b$ 12. 5 13. $2xy$ 14. $1m^2n^4$ 15. $8rs$ 16. $3x$
17. t^2v 18. $6c^5d^4f^4$ 19. 1 20. $10pqr$
21. Possible answers: $6a^3b^2$, $12a^5b^3$, $18a^4b^3$
22. 16 23. 15
24. No. 6 is not a factor of 64. GCF for 48, 54, 64 is 2.
25. Yes. The lesser exponent is a factor of the greater.
26. No. 1 is a common factor for all monomials.

10•2 Factoring Expressions

1. Work backwards and multiply each term
2. 4 3. $5p^2$ 4. 4 5. 1 6. $6s$ 7. $3x$ 8. $2xy(4x - 5)$
9. $2a^2b^2(7a^2 + 8b)$ 10. $t(t^2 + 10t + 5)$
11. $3c^3d^6(3c^2 + 4d)$ 12. $z(21z^3 + 24z^2 - 7)$
13. $2mn(2mn - 3)$
14. No. It should be $(x + 8)(x - 7)$.
15. $(p - 5)(p - 3)$ 16. $(x + 7)(x + 3)$

17. $(w - 6)(w + 5)$ 18. $(y + 3)(y + 2)$
19. $(x - 4)(x - 3)$ 20. $(m + 1)(m - 8)$
21. $4b^2 + 24b + 32$
22. Possible answer: The sign of the factor with the larger absolute value should be the same as the sign before b.
23. Possible answers: The Distributive Property is used when factoring out the GCF from two or more terms, and when factoring a trinomial into its two binomial factors.

10•3   Factoring Quadratic Expressions

1. Yes. Each quantity has a common factor of 2.
2. $3(x + 1)(x + 2)$ 3. $(7a + 1)(a + 2)$
4. $(w + 1)(5w - 1)$ 5. $(x - 1)(2x - 9)$
6. $3(n + 1)(n - 4)$ 7. $(2a + 3)(a - 2)$
8. $2(5n - 2)(n - 3)$ 9. $(11a + 4)(a - 5)$
10. $4(p - 2)(p - 9)$ 11. $5(q + 5)(q + 2)$
12. $4y$ and y or $2y$ and $2y$ 13. $6(p + 1)(p + 1)$
14. $2(5s - 1)(s + 1)$ 15. $8(x + 2)(x - 1)$
16. $3(4a - 7)(a + 2)$ 17. $(3w + 4)(2w - 1)$
18. $(6g - 1)(g - 5)$ 19. $2(4k - 1)(k - 1)$
20. $4(a + 9)(a - 1)$ 21. $(x^2 + 6x)(x - 2)$
22. $2(y^2 - y)(8y - 3)$
23. a. $(3x - 2)$ and $(x + 4)$ b. 4 ft × 8 ft × 10 ft
24. Possible answer: It is easier to factor out the 8 because that eliminates a number of factor possibilities you have to consider as you complete the problem.

10•4   Factoring Differences of Two Squares and Perfect Square Trinomials

1. Yes. Both terms are perfect squares.
2. $(3x + 1)(3x - 1)$ 3. $(5r + 6)(5r - 6)$
4. $(a + 10)(a - 10)$ 5. $(x^2 + 7)(x^2 - 7)$
6. $(12 + y)(12 - y)$ 7. $(2a^2 - 3b)(2a^2 + 3b)$
8. $(25^2 + 15)(25^2 - 15)$ 9. $(a^4 + b^2)(a^4 - b^2)$
10. $(6xy + x^2y^2)(6xy - x^2y^2)$ 11. $(8 - w^2v)(8 + w^2v)$
12. Yes. Squaring the factors results in the quadratic expression.
13. $(x - 5)^2$
14. Not a perfect square trinomial
15. $(m - 6)^2$ 16. $(2s + 3)^2$
17. Not a perfect square trinomial
18. $(2x - y)^2$
19. a. $4x^2 + 60x + 225$ b. Perfect square trinomial
20. Possible answers: $x^2 + 7x + 10$ because 10 is not a perfect square; $x^2 - 10x + 9$ because 10 is not twice the product of x and 3.

10•5   Zero Products

1. $x = -5$ or $x = 5$ 2. $m = 0$ or $m = 1$
3. $v = -5$ or $v = -6$ 4. $y = -9$ or $y + 1$
5. $z = -14$ or $z = -21$ 6. $x = 0$ or $x = 5$
7. $n = 19$ or $n = 117$ 8. $x = 0$ or $x = -3$ or $x = 4$
9. $w = 0$ or $w = -1$ 10. $m = 0$ or $m = 6$ or $m = 7$
11. $(x + 8)(x - 6) = 0$ 12. $y(y - 4)(y - 9) = 0$
13. Possible answers: Michael could have mistakenly hit 0 for the total number of weeks, resulting in multiplying all his figures by 0. He could have been using the memory and hit 0 by mistake at any point in the computation.

10•6   Solving Equations by Factoring

1. Subtract 21 from both sides.
2. $x = 2$ or $x = 1$ 3. $n = -6$ or $n = -4$
4. $y = -8$ or $y = 2$ 5. $x = \frac{3}{5}$ or $x = -1$
6. $m = 3$ or $m = 2$ 7. $x = \frac{5}{2}$ or $x = -2$
8. Cannot be solved by factoring
9. $w = \frac{3}{5}$ or $w = 2$
10. Yes. He should finish solving the problem. $t = 0$ or $t = 7$ or $t = -4$
11. $x = 0$ or $x = 3$ or $x = 4$
12. $m = 0$ or $m = -\frac{1}{2}$ or $m = 1$
13. $y = 0$ or $y = 2$ or $y = -2$
14. $n = 0$ or $n = 3$
15. $w^2 - 3w - 108 = 0$. The lower bed cannot have a negative $w = -12$ or $w = 9$ dimension. So the width is 9 and the length is 12 ft.
16. a. Possible answer: $x^2 - 8x + 12 = 0$ b. Possible answer: A solution has to make a factor 0. So, if 5 is a solution, then $x - 5$ is a factor. If -3 is a solution, then $x + 3$ is a factor.

10•7   The Quadratic Formula

1. $x \approx 1.22$ or -0.55 2. $w \approx -0.7(-0.76)$ or -5.254
3. $r = -0.5$ or -2 4. $y \approx 4.37$ or -1.37
5. $x \approx 0.17$ or 5.93 6. $s = -0.5$ or $s = -4$
7. a. $n^2 + (n + 1)^2 = (n + 2)^2$ b. $n = 3$ or -1
8. Possible answer: The quadratic formula would not give an answer because the denominator would be zero. If $a = 0$, the equation is not quadratic.

Chapter 10 Review

1. greatest common factor
2. quadratic formula
3. differences of two squares
4. 3   5. 1   6. 4   7. $3x$ 8. $3a^3b^2(6ba - 4)$

Answers   53

9. $(x + 1)(x - 9)$ 10. $(b - 9)(b + 9)$
11. $2(2y - 1)(y + 2)$ 12. $(3r - 5)(3r + 2)$
13. $(4x + 5)(4x + 5)$ 14. $(y - 7)(y + 6)$
15. $(3a - 2b)(3a + 2b)$ 16. $t = -3$ or $t = -1$
17. $x = 0$ or $x = 8$ 18. $m = -15$ or $m = 17$
19. $n = 0$ or $n = 2$ or $n = 3$ 20. $w = 1$ or $w = 11$
21. $x = 5$ or $x = -4$ 22. $r = \frac{5}{2}$ or $r = -7$
23. $y = \frac{3}{5}$ or $y = 2$ 24. $r \approx 3.58$ or 0.42
25. $x \approx -0.76$ or -5.24 26. $m \approx 0.56$ or -3.56
27. $x \approx 0.39$ or -3.9 28. 11 inches by 17 inches
29. Possible answer: Roger wants to put a fence around his rectangular garden. He wants to use the fewest number of fence posts, but wants the posts to be equally spaced. If his garden is 48 feet by 84 feet, how far apart should Roger place the fence posts?

Chapter 10 Practice Test

1. 2   2. y^2 3. $(x + 4)(x - 4)$ 4. $15r(rs + 3)$
5. $(y + 7)(y + 7)$ 6. Not possible to factor.
7. $(a^2 + b^3)(a^2 - b^3)$ 8. $4(y - 2)(y - 2)$
9. $2(3m - 1)(2m + 1)$ 10. $b^2(7a^4 - 4)$
11. $x = 0$ or $x = 3$ 12. $a = -4$ or $a = 5$
13. $w = -7$ or $w = 2$ 14. $x = \frac{1}{2}$ or $x = \frac{5}{3}$
15. $y = 4$ or $y = -3$ 16. $m = \frac{2}{5}$ or $m = 2$
17. $x \approx -0.4$ or -2.6 18. $z = 1$ or $\frac{1}{4}$
19. $n \approx -3.58$ or -0.42 20. $y \approx 2.73$ or -0.73

Chapter 10 Test

1. 6   2. $7xy$ 3. $(y + 5)(y - 5)$ 4. $9r^2s(2 - 5rs^2)$
5. $(x + 4)^2$ 6. $(a - 6)(a + 5)$ 7. $(x - 3)(x - 2)$
8. $x(2y - z)$ 9. $a(a + 12)^2$ 10. $(5c - 8)^2$
11. $y = 0$ or $y = -4$ 12. $a = 7$ or $a = -3$
13. $(y - 9)(y + 5) = 0$ so $y = 9$ or $y = -5$
14. $(6a + 5)(a - 2) = 0$ so $a = 2$ or $a = -\frac{5}{6}$
15. $(y + 9)(y - 7) = 0$ so $y = -9$ or $y = 7$
16. $(3x + 2)(4x - 3) = 0$ so $x = -\frac{2}{3}$ or $x = \frac{3}{4}$
17. $a = 5$ or $a = -8$ 18. $x = -0.76$ or $x = -5.24$
19. $x = 0.4$ or $x = -1$ 20. $x = 2.77$ or $x = -1.27$

Chapter 11 Systems of Equations

11•1 What Is a System of Equations?

1. No   2. Yes   3. Yes   4. No   5. Yes   6. Yes
7. $y = x + 4$, $x + y = 36$
8. $3m + 2c = 11$, $2m + 3c = 9$

9. Possible answer: My sister is 5 years older than I am. Her age is 1 less than double my age. How old is my sister? $y = x + 5$ and $y = 2x - 1$

11•2 Solving a System of Equations by Graphing

1. Possible answer: (1, 8); (2, 9); (3, 10)
2. (1, 1)

x	y		x	y
0	−2		0	−1
1	1		1	1
2	4		2	3

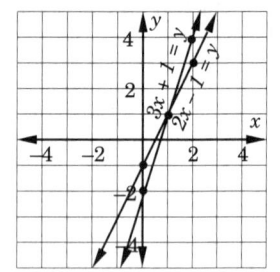

3. (1, −1);

x	y		x	y
0	−2		0	1
1	−1		1	−1
2	0		2	−3

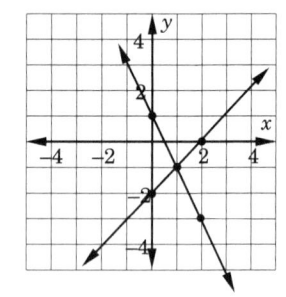

4. (3, 2);

x	y		x	y
0	5		0	−1
1	4		1	0
2	3		2	1

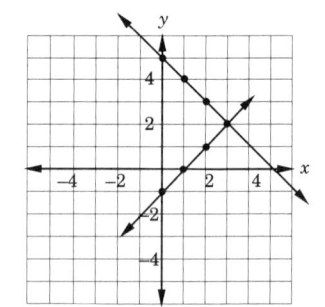

5. (−2, 1);

x	y		x	y
0	0		0	3
2	−1		1	4
4	−2		2	5

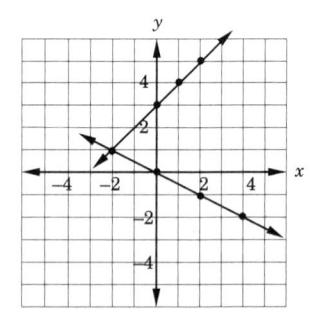

6. Possible answer: Sophie realized that the two equations named the same line, so their graphs would be the same, which means an infinite number of solutions.

7. no solution;

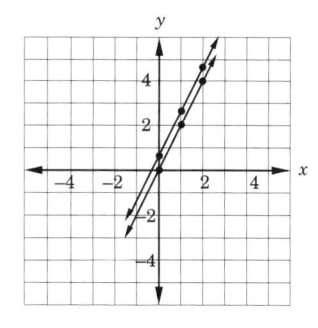

x	y
0	0
1	2
2	4

x	y
0	$\frac{1}{3}$
1	$2\frac{1}{3}$
2	$4\frac{1}{3}$

8. infinite solutions;

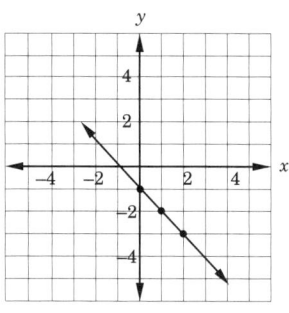

x	y
0	−1
1	−2
2	−3

x	y
0	−1
1	−2
2	−3

9. 2 hours later
10. Possible answer: Plot each line and find the point where all three or four lines intersect.

11•3   Solving a System of Equations Using Addition or Subtraction

1. No. Possible answer: Wanda solved correctly for x, but then she solved incorrectly for y, which should be −2.
2. $x = -1$ and $y = \frac{3}{2}$ 3. $x = 2$ and $y = 4$
4. $x = 9$ and $y = 12$ 5. $x = \frac{9}{2}$ and $y = \frac{15}{8}$
6. $x = 4$ and $y - \frac{1}{2}$
7. Possible answer: Neither method will solve this system of equations.
8. $x = 3$ and $y = 1$
9. $x = \frac{1}{5}$ and $y = -\frac{3}{5}$ 10. $x = 8$ and $y = 3$
11. $x = 3$ and $y = -1$
12. $0.25q + 0.10d = 14.65$ and $0.25q = 0.10d + 1.85$; 33 quarters and 64 dimes
13. $2x + y = 58$ and $6x + y$ 118; $15 for T-shirts and $28 for jeans
14. Possible answer: I have 3 more dimes than nickels. The number of dimes is 1 more than double the number of nickels. How many dimes and nickels do I have?

11•4   Solving a System of Equations Using Multiplication

1. $x = 9$ and $y = 6$ 2. $x = 7$ and $y = 17$
3. $x = 27$ and $y = 10$ 4. $x = 1$ and $y = 3$
5. $x = 6$ and $y = 5$ 6. $x = \frac{32}{7}$ and $y = -\frac{26}{7}$

7. first by 2; second by 3   8. $x = 6$ and $y = 7$
9. $x = -\frac{22}{13}$ and $y = \frac{29}{13}$ 10. $x = -4$ and $y = -2$
11. $x = -\frac{3}{11}$ and $y = \frac{64}{11}$ 12. $x = 1$ and $y = 3$
13. $x = 2$ and $y = 5$
14. 75 pounds ValleyVanilla, 25 pounds MountainMist
15. Possible answer: Solving a system of equations is one method for getting an answer to a question or problem, provided that the problem can be represented this way.

11•5   Solving a System of Equations Using Substitution

1. Possible answer: Replace the y in the second equation with $4x$.
2. $x = 2$ and $y = 1$ 3. $x = 4$ and $y = -2$
4. $x = \frac{20}{3}$ and $y = \frac{5}{3}$ 5. $x = 2$ and $y = 4$
6. $x = -2$ and $y = -1$ 7. $x = 1$ and $y = 7$
8. They are both correct ways to solve the problem.
9. $x = -3$, $y = -5\frac{1}{2}$,
10. $x = 5$, $y = 10$,
11. width is 80 feet; length is 240 feet
12. Possible answer: Addition or subtraction, with multiplication. This method will work for any system. Substitution is useful only when one variable is easily replaced, and graphing can make it difficult to read fractional answers.

Chapter 11 Review

1. system of equations   2. Solution set
3. graphing; addition or subtraction with multiplication; substitution
4. Yes   5. Yes   6. No   7. Yes
8. (2, 3)

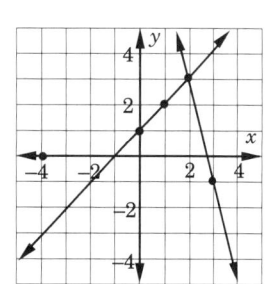

9. (1, 4)

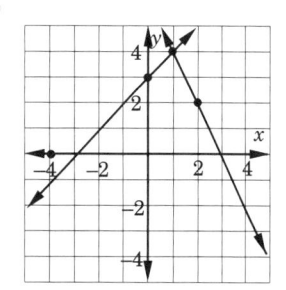

10. $x = \frac{5}{3}$ and $y = \frac{10}{3}$
11. $x = \frac{5}{2}$ and $y = 2$
12. $x = 6$ and $y = 7$
13. $x = -\frac{102}{29}$ and $y = -\frac{50}{29}$
14. $x = 12$ and $y = 8$
15. $x = -\frac{22}{13}$ and $y = \frac{29}{13}$
16. 4 and 8
17. plants are $2; seeds are $1
18. 10 quarters
19. Robin has 15 CDs; Nigel has 33 CDs
20. Answers will vary.

Chapter 11 Practice Test

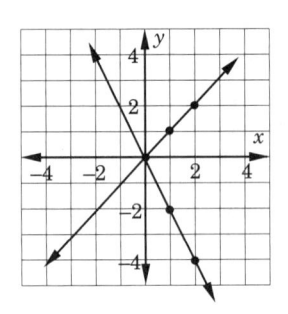

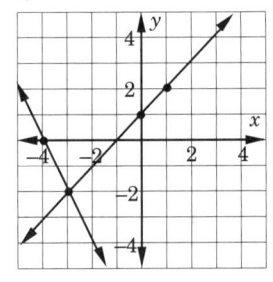

1. Yes
2. (0, 0)
3. (−3, −2)

4. $x = -6$ and $y = -6$ 5. $x = 4$ and $y = -1$
6. $x = 2$ and $y = 1$ 7. $x = 7$ and $y = 11$
8. $x = -4$ and $y = 10$ 9. $x = 9$ and $y = 2$
10. $3 for pair of socks; $45 for gym shoes

Chapter 11 Test

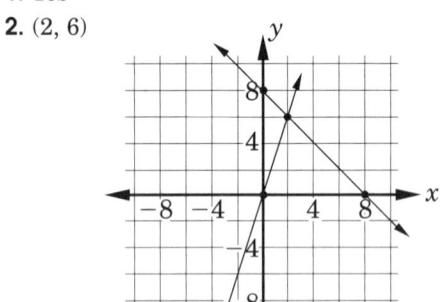

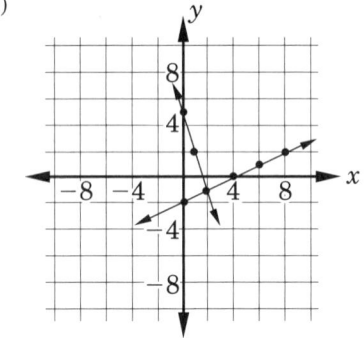

1. Yes
2. (2, 6)
3. (2, −1)

4. $x = 12$ and $y = 15$ 5. $x = 3$ and $y = 2$
6. $a = 2$ and $b = 3$ 7. $x = -3$ and $y = 2$
8. $x = 4$ and $y = 7$ 9. $x = -2$ and $y = 2$
10. 20 blue and 15 red

56   Answers